KV-370-156

WELDING SCIENCE AND METALLURGY

Written as a sequel to the same authors' *Welding Craft Science*, this book has been designed to cover the science content of Part II of the City and Guilds of London Institute new Welding Craft Practice (323) course and the old Welding (74) course. In addition, the science sections of the equivalent welding syllabuses drawn up by the U.L.C.I. and the other regional examining bodies are fully covered, as is the science of welding in the City and Guilds Heating and Ventilation schemes (256 and 257). Once again, the clear and logical way in which the subject is presented will recommend the book to students both in technical colleges and in industrial craft training schemes.

This book is particularly intended to impart a precise knowledge of the metallurgical problems encountered during and after welding of differing materials, and will thus be of interest to engineers and draughtsmen in general who are without specialized knowledge of welding processes. A special effort has been made to cover the more difficult and sparsely described materials: for example, mode of solidification, effects of cooling rates by different processes, simple stress concentrations applied to weld defects, change of type of fracture with temperature, simple mechanism of intercrystalline corrosion, welding clad steels, copper and heat-treatable aluminium welding, types of wear, plasma arc cutting, and joining of dissimilar materials.

Both authors are examiners in welding and are on the staff of Bolton Technical College.

By the same authors:
Welding Craft Science

Welding Science and Metallurgy

by **KENNETH LEAKE**

Member of the American Welding Society.
Lecturer in charge of Welding at Bolton Technical College.
Local Examiner in Welding for the City and Guilds of London
Institute and the Lancashire and Cheshire Institutes.
Member of the Lancashire and Cheshire
Institutes Welding Committee.

and **NORMAN J. HENTHORNE**

Member of the American Welding Society.
Lecturer in Welding at Bolton Technical College.
Local Examiner in Welding for the City and Guilds of London
Institute and the Lancashire and Cheshire Institutes.
Member of the Lancashire and Cheshire Institutes Welding
Committee.

Donagh O'Malley,
R. T. C.
Letterkenny

CASSELL · LONDON

B/671.52

CASSELL & COMPANY LTD
35 *Red Lion Square*, *London WC*1
Melbourne, Sydney, Toronto
Johannesburg, Auckland

© Cassell & Co. Ltd, 1968
First published 1968

S.B.N. 304 92835 6

Phototypeset by BAS Printers Limited, Wallop, Hampshire
Printed in Great Britain by
Ebenezer Baylis and Son Limited,
The Trinity Press, Worcester, and London
F.1268

Contents

Bibliography

Riddihough, M., *Hardfacing by Welding:* pub. by Deloro Stellite Ltd.
Metals and how to weld them: James F. Lincoln Welding Foundation.
Mond Nickel Publications.
Copper Development Assn Publications.
Welding Handbook, vol. 1–5: American Welding Society.
Henry, O. & Clausen, *Welding Metallurgy:* pub. by the American Welding Society.
American Welding Society Journals.
Rockwell Publications.

Preface

This book is intended to meet the need for a comprehensive and up-to-date text book in welding science. Welding practice and technology have advanced considerably over the past years. The result of this is that many branches of welding science need specialization. It has been the aim throughout the writing of this book to keep this in mind and apply science as an aid to the adequate treatment of welding practice and technology.

It has also been the intention to present in a simple form the principles of science applied to welding. The book will be found to be particularly suitable for students in the Advanced Craft Courses of Welding and also Welding Technicians Courses. It will also be found useful to those who are studying courses which include welding as an ancillary subject—steel fabricators, heating and ventilating engineers, plumbers, motor vehicle mechanics, etc.

In industry, welding engineers, foremen, draughtsmen who are faced with the problem of deciding on the type of process and filler material for a given application, will find that this book is of use to them.

The contents give a simply explained treatment of alloying, the metallurgy of iron and steel as applied to welding, including weaknesses during welding and the various forms of heat treatment. The alloy, heat resisting and stainless steels are discussed along with the various forms of cast-iron, also the non-ferrous metals copper, aluminium and their alloys along with their weldability. A suitable chapter on hardsurfacing is included and a final chapter on the various forms of cutting, ranging from oxy-acetylene to laser cutting.

It is believed that the welder should become acquainted with the reasons why certain processes are preferable to others for the welding of various materials, and why metallurgical changes may occur. During the course of writing this book, the authors have tried to relate their own practical experiences to the science involved. A study of what occurs under the source of welding heat and how a material will react when welding is simply explained. Each individual metal or alloy is dealt with separately, and every attempt is made to explain what effect the welding heat will have on the properties of the material.

Acknowledgements are due to the following for permission to use information: Deloro Stellite Ltd., the Copper Development Assn., the Lincoln Foundation for their permission to reprint extracts from 'Metals and how to weld them', the Mond Nickel Company, Rockwell Ltd.

Finally our thanks are given to our artist Mr J. Critchley who has cleverly illustrated the text throughout the book.

K. Leake
N. J. Henthorne

Copies of British Standards referred to in this book may be obtained from: British Standards Institution, 2 Park Street, London W.1.

CHAPTER 1

Principles of Alloying

When carrying out practical welding, it is rare to weld metals which are *absolutely* in their pure state, that is, containing no other alloying materials. Even metals regarded as commercially pure usually contain either small amounts of impurities or small amounts of other elements deliberately added for various reasons; for example, to assist in deoxidizing, to allow the metal to flow more easily, or to improve the mechanical or physical properties of the metal concerned.

It is well worth while investigating the very desirable properties which may be obtained by alloying. If we take iron alone, it is not considered strong enough or hard enough for practical purposes. Moreover, if we wanted it in its absolutely *pure* state it would also be most expensive. The slightest amount of carbon immediately improves its strength and hardness. If the carbon content is below 0·3% the metal is known as mild steel. This is quite ductile and there are no special difficulties during welding. If the carbon content is increased slightly, to 0·5%, there is an increase in the strength and hardness of the metal, but it is no longer ductile, it becomes brittle, and it is more difficult to weld. It can be seen that the addition of only a small percentage of carbon controls the difference between a ductile material which is readily weldable and a brittle material which can only be welded if great care is taken.

Another well-known metal, aluminium, has poor mechanical properties, its tensile strength being about 6 tons per square inch; it is also quite soft. Its properties can be greatly improved by alloying, the most used of alloying elements being magnesium, silicon, zinc, copper, manganese, etc. With 4% copper added, the tensile strength of the alloy formed is equal to mild steel, and its hardness increased to over 100 on the Brinell hardness scale.

If consideration is given to adding alloying elements to copper, a wide range of brasses and bronzes can be produced, each with their own special applications. In certain cases, the amount of the alloying element added is quite high, e.g., up to 45% zinc. This gives an alloy which has a high tensile strength and is corrosion resisting.

When welding commercially pure copper which has been deoxidized, success is brought about through the alloying element phosphorus, which is added to the copper during manufacture. The percentage of phosphorus added is between 0·01 and 0·04%; despite this being quite a small amount, it is very effective, preventing any copper oxide from forming.

The behaviour of the different alloying elements cannot be predicted unless the fundamental principles of why alloying elements are added, what form they will take in the material being welded, and how they will react when heated and cooled, are thoroughly understood. Fortunately, the study of alloying is quite straightforward. It is worth while grasping the basic ideas of solubility, mixtures, compounds, solid solutions, etc., to get a complete 'picture' of the importance of alloying to the engineering industry.

Solutions

The most well-known solutions of all simply consist of water and some form of salt. If we take a pint of water, place it in a glass beaker and add one teaspoonful of household salt (sodium chlorate) and stir it up, the salt will go into solution with the water. When holding this solution in front of a light, no trace of the salt is seen. The important thing to remember is that it is a perfectly homogeneous *mixture*. It has not formed a *compound* and, if the water is evaporated off in the form of steam, the salt will be seen at the bottom of the beaker in its original form.

Instead of evaporating the water off, if more salt is added to the water, a stage will be reached when the water is incapable of dissolving any more salt, providing that the temperature remains constant. The water-salt solution is then said to be *saturated*, and is known as a saturated liquid solution. If the temperature of the liquid is increased, more salt will dissolve into it. This means that there is an increase in the solubility with a rise in temperature.

However, if the temperature is lowered slowly, the salt in excess of that which causes saturation will precipitate out again.

Another familiar example of solubility is to be found when dealing with acetylene gas in cylinders. This is commonly known as 'dissolved acetylene', and refers to the fact that the acetylene is stored under pressure in a liquid known as acetone, which is capable of dissolving, or taking into solution, twenty-five times its own volume of acetylene gas per atmosphere of pressure. When welding, acetylene gas is consumed, causing a small reduction of pressure in the cylinder. In turn, this causes more acetylene gas to precipitate out of the solution which is formed when acetylene is dissolved in acetone under pressure.

Having established that a solution is a perfectly homogeneous mixture, it is now possible to consider the solution which may remain after the solidification of metallic substances. This is known as a metallic solid solution.

Metallic solid solutions

When a metal solidifies, the atoms of that metal arrange themselves into a definite pattern or atomic structure. When viewing a metal with a microscope, if we consider one crystal out of the many thousands so formed, we will be dealing with millions of atoms. In a pure metal each atom has a definite distance away from the next atom. It is very important to remember that each individual

Fig. 1. 'Body centred cubic' atomic arrangement.

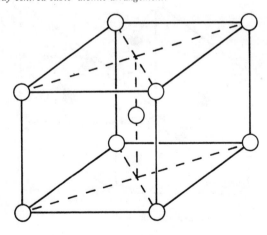

3

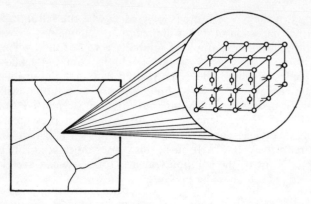

Fig. 2. Showing arrangement of atoms if high enough magnification were possible.

metal has its own individual distance between its atoms. We could never hope to see any atoms with a microscope because they are extremely small, but each separate element has its own particular size of atom and use can be made of this when alloying.

Iron, chromium, tungsten and molybdenum, when cold, have what is known as 'body centred cubic' atomic arrangement, i.e., one atom at each corner and one in the middle. (See Figs. 1 and 2. The lines on Fig. 1 are added for guidance only, they do not, in fact, exist.)

The size of the crystal is formed by more and more atoms taking up a definite pattern.

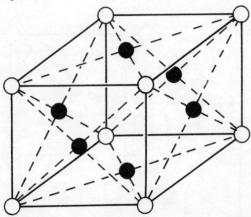

Fig. 3. 'Face centred cubic' atomic arrangement.

4

If copper, nickel, aluminium or lead solidify, they will take up a 'face centred cubic' atomic formation. (See Fig. 3.) This refers to the fact that the atoms form themselves into groups of fourteen, with one atom at each corner and one in the centre of each face. Metallic solid solutions may be conveniently divided into two groups, these being known as 'substitutional' and 'interstitial' solid solutions. In both cases, the microscope will only show crystals of a homogeneous solution of one element dissolved in a metal. It is not possible to detect the individual elements with the microscope; the crystals will have the appearance of a pure metal.

Substitutional solid solution

This is formed when two metals, which are soluble in one another in the liquid state, remain dissolved in one another in the solid state. For this to happen, the atoms of one metal must be capable of fitting into the atomic structure pattern of the other metal. This occurs if the size of the alloying element atoms are within 15% of one another. The manner in which they occupy positions is by substitution of the atomic arrangement. If the alloy was solidifying, taking on a body centred cubic arrangement, an end view of this would appear as in Fig. 4, the black atoms representing one metal and the white ones representing the substitute metal. When the size of the added alloying element atom is larger than the base metal atoms, in taking up their positions in the atomic structure, they will occupy more room. The result of this is that

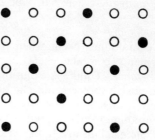

Fig. 4. End view of substitutional solid solution.

there is less space between the atoms. This causes an increase in *hardness* and *strength*, but usually a small loss in *ductility*. The metals copper and nickel are typical examples of complete solubility in the liquid and the solid state.

Interstitial solid solution

The word 'interstitial' comes from the Latin word *intersistere* which means 'in between', in this case referring to the spaces existing between the atoms. This type of solid solution is formed when the atoms of the added elements are very small compared with the atoms of the base metal. The smaller size atoms are often

5

elements which exist in a gaseous state at room temperature. For example, nitrogen is the element with the smallest size of atom, and other gases, such as oxygen and hydrogen, also have small atoms. From this it can be understood that the small atoms of the gases mentioned can exist in the spaces between the larger atoms of a metal. Carbon atoms are also very small and may exist between the spaces provided by the atomic arrangement of a solid metal such as iron. (See Fig. 5.) In the case of pure iron, the amount of carbon that it can hold in solution is only 0·006% at 0°C. Should carbon be present above this amount, it will react *chemically* with the iron, forming a compound. However, by discreet use of adding further alloying elements which allow for more carbon to be kept in an interstitial solid solution, the important industrial alloy of austenitic manganese steel is produced.

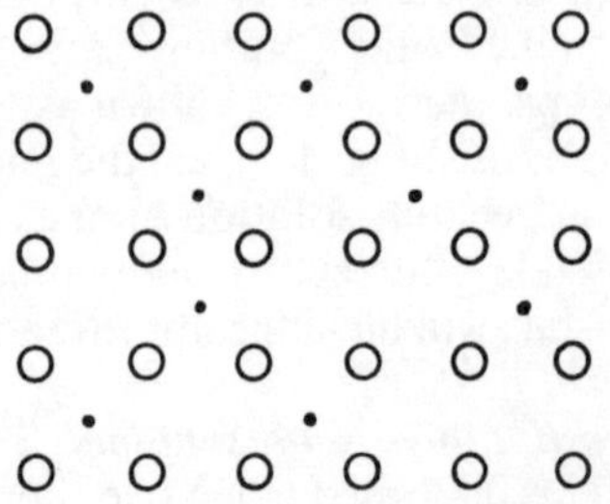

Fig. 5. The interstitial solid solution.

Compounds

Compounds occur when the atoms of two or more individual elements combine chemically at their surface forming another substance. Unlike a solution, the compound formed does not necessarily possess the characteristics of the individual elements from which it is composed. It is usually difficult to separate a compound back to its individual elements, but it can be done, often by chemical means. In the welding industry, compounds have many uses, including the following:

(1) Temperature-indicating crayons rely upon a definite melting point which is a characteristic of a compound.

(2) When oxy-acetylene welding certain materials, fluxes are required to dissolve oxides which may be present, or that form on the surface during welding. Fluxes are usually composed of compounds, and turn liquid at a temperature below the melting point of the material. From this temperature upwards they combine with the oxides.

(3) In alloys, the formation of a compound by the addition of small amounts of various elements may increase the hardness and strength of the alloy. Examples of this can be illustrated

by adding carbon to iron, chromium or tungsten, forming, respectively, iron carbide, chromium carbide and tungsten.

The intermetallic compound

This is a chemical combination of two metals. The compound so formed has definite characteristics, such as high hardness, brittleness, a definite melting point. It is usually easily recognizable when viewed through a microscope because it is a different colour from either of the materials from which it is formed. The intermetallic compound is not used alone, but forms many useful alloys when used in conjunction with many metallic solutions. Duralumin is an alloy which contains approximately 4% copper. This copper first forms a solution with aluminium and then an intermetallic compound ($CuAl_2$). Correctly balanced, this alloy possesses the strength and hardness of mild steel, but is much lighter.

Other examples of intermetallic compounds include copper-tin (Cu_3Sn), iron-tungsten (Fe_2W_3) and magnesium-tin (Mg_2Sn).

The effects of alloying may be considered as follows:

(1) Alloying elements may form substitutional or interstitial solid solutions, chemical compounds or remain mixed up with the solid alloy. The effect of alloying elements is often to improve the strength and hardness of a material or to improve its corrosion resistance.

(2) Some elements are added to the alloy to assist in deoxidizing, thus keeping the alloy free from undesirable oxides which may weaken the material and possibly make it less corrosion resisting.

(3) Often an alloying element is added for the purpose of making the material flow more easily when in the liquid state, use being made of lowering the melting point.

(4) An alloy may freeze over a range of temperature, or, in exceptional cases, at one temperature. Considerable use is made of both modes of freezing by filler-rod manufacturers when producing suitable materials for soldering, brazing and welding.

(5) When alloying elements are added to a metal, the thermal and electrical conductivity of the metal is always lowered. If copper is taken as an example, it will be found that in the pure state it is an excellent conductor of heat and electricity. Conduction plays an important part when welding. The difficulties which are associated with the welding of copper are mainly due to the high rate of conduction. If zinc is

Welding Science and Metallurgy

added to copper, forming brass, the conduction rate is reduced and the heat balance between that lost by conduction and that required for welding is easier to maintain. In general, it is easier to weld a copper alloy than pure copper.

Unfamiliar terms used in this chapter

Alloying: The mixing of metals together.

Deoxidized: A material containing an element which prevents oxygen from having detrimental effects is said to be deoxidized.

Physical properties: These may be plasticity, expansion and contraction, conduction rate, melting point, fusability, etc.

Solubility: Capable of a solution being formed.

Evaporating: Turning into vapour, i.e., changing water into steam.

Saturated: To be dissolved to a maximum amount.

Precipitate: To reject or throw out.

Homogeneous: Composed of similar parts.

Austenitic: A microstructure of a solid solution of iron, when the structure is face centred cubic.

Chapter review questions

What are the advantages of alloying metals?
Explain the characteristics of an intermetallic compound.
Describe two different types of solid solutions.
What is an intermetallic compound?
Explain the effect of adding 4 % copper to aluminium.
Why is a small percentage of phosphorous added to copper?
What is a solid solution?
List four uses of compounds advantageous to the welding industry.

CHAPTER 2

The Influence of Carbon on Iron

Pure iron

Iron is a remarkable metal. When cooled from the molten state, several changes take place, occurring while the iron is in the solid state. Molten iron freezes at 1 530°C, the temperature then drops steadily until 1 405°C is reached. The iron then remains at this temperature for a short time, indicating that a change is taking place in the solid iron; this particular change, whilst noted by metallurgists, is of little importance.

On further cooling, the temperature remains constant for a short time at 910°C; a change now takes place internally and it is very important. The iron atoms are rearranging themselves internally to form another type of crystalline structure. Above 910°C the iron atoms are arranged in a face centred cubic formation. This is known as gamma iron. (See Fig. 6.) The reason that the temperature remains constant at 910°C is that heat is given off during the atomic rearrangement. The temperature remains constant until all the atoms have taken up the body centred cubic formation. The temperature at which the change in crystalline structure takes place is known as the transformation temperature. The body centred cubic arrangement of atoms below 910°C is now known as alpha iron.

On further cooling to 760°C another change takes place; this is the return of the magnetic properties to the iron, and is known as the Curie point.

If a piece of iron is heated, the transformation takes place in a reverse manner. On reaching the various change points on heating, heat is absorbed instead of being given out, as is the case when iron

B

9

is cooled. Hence, the transformation points when heating are slightly higher than when cooling.

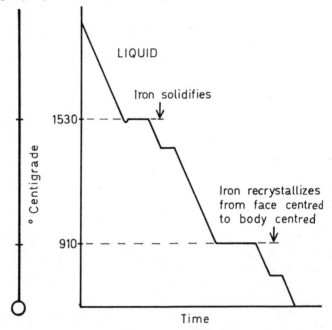

Fig. 6. Changes which occur when pure iron is cooled.

Addition of carbon to iron

Iron becomes known as steel when carbon is added. It is interesting to consider the effects of adding carbon to iron. Pure iron has a definite freezing temperature. The influence of the carbon is to cause the steel to freeze over a range of temperature. The more carbon there is in solution with the iron, the lower the range of temperature at which freezing will take place.

Iron with a body centred cubic type of atomic structure can only hold a trace of carbon in solution with it (so little that it may be completely ignored). When iron has a face centred cubic atomic arrangement, it is capable of holding a considerable amount of carbon, forming a solid solution. The solid solution of carbon in iron is known as austenite. Under the microscope, austenite has the appearance of a pure metal, rather than an alloy. (See Fig. 7.)

Because a considerable amount of carbon can be held in solution with iron when the structure is austenite, it is possible, by controlling the cooling rates, to regroup the carbon in the final structure at room temperature. This will give desirable mechanical properties. When iron or steel is slow cooled or cooled under equilibrium conditions, the

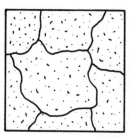

Fig. 7. Austenite × 1 000.

structure would appear under the microscope as in Fig. 8. If the structure of pure iron in Fig 8(a) is considered, many grains of pure metal are visible, each being an equiaxed grain with its own grain boundary. If a small amount of carbon is added,

(a) <u>Pure Iron</u>
(Ferrite)

(d) <u>0·83% Carbon</u>
All Pearlite

(b) <u>0·25% Carbon</u>
Ferrite and
Pearlite

(e) <u>1·2% Carbon</u>
Cementite
surrounding
Pearlite

(c) <u>0·6% Carbon</u>
Ferrite and
more Pearlite

(f) <u>1·7% Carbon</u>
More Cementite
surrounding the
Pearlite

Fig. 8. The changes in the microstructure when carbon is added to iron.

that is, up to 0·006% at room temperature, it goes into solution with the iron and is not visible under the microscope. Pure iron and iron containing this small amount of carbon are known as ferrite. Ferrite has the following properties:

(a) It is soft and ductile.

(b) It has an ultimate tensile strength of around 20 tons per square inch.

(c) It can be readily hot or cold worked.

When carbon is added to iron in excess of 0·006%, a chemical compound is formed, known as carbide of iron, iron carbide or cementite. This is formed by three atoms of iron combining with one atom of carbon (Fe_3C). The main feature of this compound is its hardness. In steels, the presence of cementite increases the hardness and the tensile strength of the material; the more cementite, the greater the increase. Steel containing a carbon content of up to 0·3% is known as mild steel, and has a tensile strength of around 25 tons per square inch and a hardness value of 109 Brinell. It is quite ductile and is readily welded by most welding processes. In Fig. 8(b) the microstructure of a steel containing 0·25% carbon is shown; this is typical of the structure of mild steel. In addition to the ferrite grains, there are grains that appear in black and white layers, resembling the edge of a piece of plywood. These layers contain iron carbide (cementite) and ferrite which are sandwiched alternately between each other. The name given to these areas is pearlite. The amount of carbon in the pearlite *never* varies, it is always 0·83%. The effect of pearlite is to give the steel an increase in tensile strength and hardness, but there is a loss of ductility. Pearlite is distributed evenly throughout the microstructure.

A further increase in the carbon content is shown in Fig. 8(c) by the presence of more pearlite. There is a further increase in the tensile strength and hardness but a notable decrease in the ductility of the material. Because of this the welding difficulties will increase. If a material for welding is not ductile, it is brittle, with an increased tendency to crack.

If the carbon content is increased to 0·83%, the entire structure will be composed of ferrite and cementite in layers, hence pearlite throughout.

If the carbon content is increased to 1·2%, it can be seen from Fig. 8(e) that the excess of cementite appears around the pearlite. The reason for this is that there is now too much carbon to form pearlite and all the carbon above 0·83% is distributed at the grain boundaries. The full range of carbon in steel is up to 1·7%, this is shown in Fig. 8(*f*). Compare the structure of a 1·7% carbon steel with the structure of a 1·2% carbon steel; it will be noted that there is a greater amount of cementite around the pearlite on the 1·7% carbon steel.

The construction of the iron-carbon thermal equilibrium diagram relating to steels

It is now possible to consider what will occur when a piece of steel is heated and cooled slowly. To understand this correctly, it is necessary to have some knowledge of what happens to the steel internally. If the steel is heated or cooled there are some important differences in the sequence of changes compared with the changes described when considering pure iron. Because carbon has been added to the iron the internal changes in the atomic structure now take place over a temperature range. (The exception to this being when the carbon content is 0·83%.) This range alters with the amount of carbon in the steel. The purpose of the iron-carbon thermal equilibrium diagram is to illustrate when the internal changes occur.

Fig. 9. The change in structure of pure iron recorded on the hypothetical thermometer.

In order to construct this diagram it is necessary to imagine that we could use thermometers for measuring extremely high temperatures. These will be referred to as hypothetical thermometers.

In putting the hypothetical thermometer to use, it can be seen that the pure iron shows one internal structural change on it, the temperature at which this occurs is 910°C. If the heating curve is taken of all compositions from pure iron up to 1·7% carbon content in the steel, the temperature at which the internal changes take place may be recorded on the thermometer. By placing the thermometers in order of carbon content from 0% carbon, which is

13

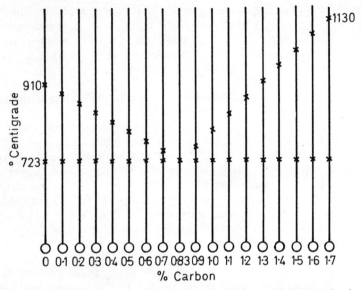

Fig. 10. The various ranges of temperature between which internal changes take place.

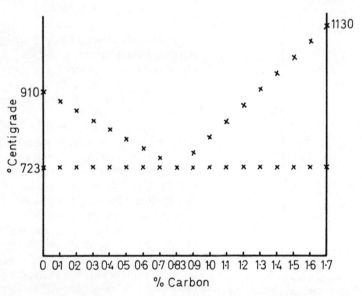

Fig. 11. The temperature ranges used for equilibrium diagram construction.

14

pure iron, to 1·7% carbon, equally spaced apart, the iron-carbon diagram for steels may be plotted. The distance spacing on the vertical scale represents temperature in degrees Centigrade. The carbon content is indicated on the horizontal base line.

By use of connecting lines the diagram may be completed. This can be seen in Fig. 12.

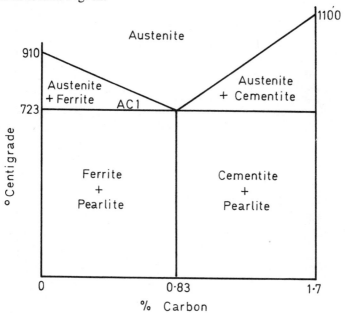

Fig. 12. The iron-carbon equilibrium diagram.

The various areas on the diagram have been labelled. These indicate the name of the constituents which are seen by looking through a microscope. It should be remembered that this diagram has been constructed from the temperatures originally obtained from heating curves. The lower horizontal line at 723°C is known as the A.C.1. line. The other line is from 910°C, joining the A.C.1. at 0·83% carbon, and is known as the A.C.3. line. Between the A.C.1. and the A.C.3. lines is the transformation range, the iron existing in both the face centred and body centred cubic atomic formation.

If the diagram was constructed with the aid of cooling curves instead of heating curves, then the lines would be referred to as A.R.1. and A.R.3. These lines appear 30°C below the A.C. lines.

15

The complete diagram for steels is shown in Fig. 13. The micro-structures of the various areas depending upon temperature and carbon content are shown. An additional vertical line is added from 0·83% carbon at 0°C, meeting the A.C.1. at the same point where the A.C.3. merges with the A.C.1. line. This denotes that the composition is all pearlite.

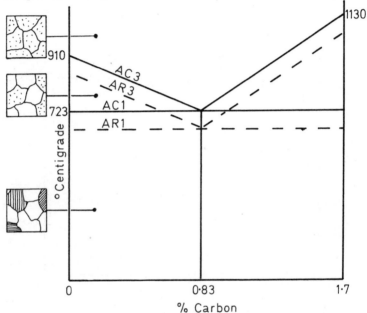

Fig. 13. The iron-carbon equilibrium diagram for steels, with added microstructures.

It is now possible to consider the various changes which may occur when steels are either heated or slowly cooled through the transformation range.

Heating steel with a carbon content below 0·83%
    A steel containing less than 0·83% carbon has a structure of ferrite and pearlite at room temperature. If the steel is slowly and uniformly heated, at a temperature of 723°C, which is the A.C.1., the pearlite changes to austenite. Whilst this internal change is taking place, heat is absorbed, there being no increase in the temperature. When the temperature continues to rise, the austenite gradually absorbs the ferrite grains, until the A.C.3. line is reached.

16

At this point the entire structure is austenite. This means that a mixture of ferrite and pearlite has changed to a solid solution, occurring while the steel is in the *solid* state.

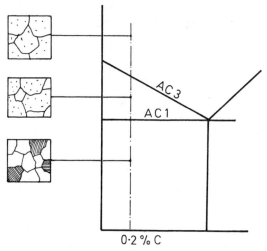

Fig. 14. Heating of steel containing 0·2 carbon.

Heating of steel containing 0·83% carbon

The structure of steel of this composition is entirely pearlite at room temperature. On heating to the A.C.1. line, which in this case merges with the A.C.3. line, the temperature remains constant while the pearlite changes to austenite. Once the structure is composed entirely of austenite, the temperature will continue to rise.

Heating of a steel containing above 0·83% carbon

At room temperature the structure is pearlite and cementite. When heated to the A.C.1. line, the pearlite changes to austenite, the temperature remaining constant until this is complete. On further heating, the austenite slowly absorbs the cementite until the structure becomes entirely austenite at the A.C.3. temperature.

Eutectoid reaction

The eutectoid reaction may be simply defined as the breakdown of a solid solution into a solid mixture. This occurs at a fixed temperature and there is a definite composition of the materials capable of this reaction.

Welding Science and Metallurgy

The eutectoid reaction occurs when a piece of steel is slowly cooled. In the case of steels, the solid solution is austenite and the solid mixture into which it breaks down is ferrite and cementite, i.e., pearlite. When cooling steel of *any* carbon content, any austenite remaining at the A.R.1. line temperature will contain 0·83% carbon. At this temperature and composition the eutectoid reaction occurs. It should be realized that this reaction occurs every time a piece of steel is slowly cooled. (See Fig. 13.)

Slow cooling of steel containing below 0·83% carbon

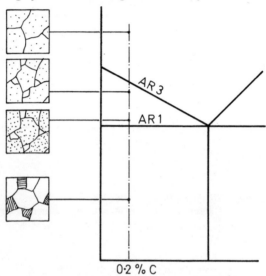

Fig. 15. Microstructures when cooling steel containing 0·2% carbon.

Steel of this composition is a solid solution at a temperature above the transformation range. All the carbon is dissolved in the iron. On slowly cooling the solid solution of austenite to the A.R.3. line, it begins to change to the body centred cubic structure of ferrite. This results in grains of ferrite precipitating out of the austenite solid solution. As the ferrite precipitates out, the austenite readily takes in the excess carbon atoms which cannot remain in the newly formed body centred ferrite, the remaining austenite becoming enriched in carbon. This continues as the temperature falls, more austenite transforms and the ferrite grains become larger and more numerous. When the A.R.1. line is reached, the austenite remaining now has a carbon content of 0·83%. This is the

18

The Influence of Carbon on Iron

eutectoid reaction temperature and the austenite changes to pearlite. The final structure is composed of grains of ferrite and pearlite.

Slow cooling of steel of 0·83% carbon composition
The steel is of eutectoid composition. This means that there is a definite transformation temperature. Above this temperature, the structure is austenite and, when slowly cooled, it will change to pearlite. This is a direct change, there being no precipitation of any other constituent. The final structure is pearlite.

Slow cooling of steel containing above 0·83% carbon
When the austenite is slowly cooled to the A.R.3. line, the solubility of carbon in the austenite is lowered so that it cannot hold all the carbon in solution. Some of the carbon is rejected and combines with the iron to form cementite. As the temperature lowers, more cementite is formed. Upon cooling to the A.R.1. line, the remaining austenite will contain 0·83% carbon. At this temperature it transforms to pearlite and the final structure is of cementite and pearlite.

It should be realized that both low carbon and high carbon steels will form austenite, having the same amount of carbon at the A.R.1. line when cooling. This is because a steel containing less than 0·83% carbon forms ferrite upon cooling through the transformation range. As the ferrite precipitates out, the carbon content of the remaining austenite will increase. The high carbon steel forms cementite between the transformation range, using up carbon and thus lowering the carbon content of the remaining austenite until, at the A.R.1. line, the carbon content will be 0·83%. The austenite *always* has a carbon content of 0·83% when at the A.R.1. line.

Quenching of carbon steels
Rapid cooling of steels from above the transformation range, when they are in the austenitic condition, may be carried out in water or oil. This is known as quenching and has the effect of not giving the steel time to carry out its internal changes in the transformation range. This means that the final structure on cooling will be entirely different to what it would have been if the steel had been slow cooled. The carbon does not have time to react as usual and pearlite does not form if the cooling rate is fast. The rapid cooling rate tends to undercool the austenite. The outcome of this is that

19

the face centred cubic atomic formation is still present at quite low temperatures and this structure is capable of holding carbon. However, on reaching the temperature range between 350°C and 150°C, the undercooled austenite *must* change its face centred cubic structure to the body centred cubic structure. This newly formed structure has very little room for carbon, and it should also be appreciated that the steel has contracted considerably on cooling. Therefore the carbon atoms are trapped in the newly formed structure and cause a distortion or a bulging of the atomic arrangement. This gives a tightening up of the atoms and, if the quench rate is fast enough, the hardest structure in steel is formed, known as martensite. (See Fig. 16.)

The higher the carbon content, the harder and more brittle is the steel. When the higher carbon steels are quenched, more carbon is trapped in the body centred arrangement. This accounts for the increased brittleness. In mild

Fig. 16. Martensite × 100.

steel, the carbon content is not high enough to give appreciable hardness, but there is a considerable loss of ductility.

Unfamiliar terms used in this chapter

Crystalline: In the forms of crystals.

Solution: A substance with something dissolved in it.

Mechanical Properties: These are tensile strength, compressive strength, elasticity, ductility, brittleness, hardness, etc.

Equilibrium: In a state of balance. Equilibrium diagrams of metals refer to metals which have cooled very slowly.

Ultimate tensile strength: When a piece of metal is pulled apart, the maximum load it can withstand is known as the ultimate tensile strength.

A.C.1.: The lower transformation temperature and the point where there is an internal change started on heating a piece of steel.

A.C.3.: This is the higher transformation temperature. Upon heating to this temperature, a piece of steel becomes completely austenite.

Chapter review questions

What is meant by 'the transformation range'?

How is austenite formed?

Name the hardest structure formed in steel.

Explain the changes which occur upon heating a steel containing 0·83% carbon.

Name the changes which occur on heating a 0·4% carbon steel to 1 000°C.

What is gamma iron?

At what temperature would a eutectoid reaction take place on slow cooling?

What effect does carbon have on the mechanical properties of iron?

Why are thermometers, used in the construction of the iron carbon equilibrium diagram, regarded as hypothetical?

What type of atomic structure would be found within the critical range?

CHAPTER 3

The Heat Treatment of Steels

All types of plain carbon steels are capable of heat treatment in some form or other. Regardless of whether the carbon content is as low as 0·15% or as high as 1·7%, they are capable of being annealed, normalized, stress relieved or hardened. It is necessary to classify the difference in the carbon content of the steels in order to be able to assess how they will react when heat treated.

Mild steel

The carbon content of mild steel may be up to 0·3%. This means that if the material is quenched from above the upper transformation temperature, there is insufficient carbon to result in any appreciable increase in hardness. It is possible to increase the hardness of the surface, and therefore its resistance to wear, by adding carbon to it. This may be done whilst the steel is in the solid state, at a temperature above the upper transformation point and is known as case hardening.

While mild steel is not regarded as a hardenable material, care should be taken, after welding, not to quench an article from a temperature within the transformation range. The reason for this is that there is a chance that the carbon may be grouped in small areas and not distributed evenly throughout the material. This means that on rapid cooling, i.e., by quenching, the carbon cannot form the desirable pearlitic structure which gives mild steel its characteristic toughness. The impact value will be drastically reduced, therefore quenching of mild steel articles after welding is to be deplored.

After welding has been carried out, residual stresses may be present through the uneven rate of contraction throughout the material. The effects of residual stresses may result in premature failure of the welded component if they are not removed by correct heat treatment, i.e., annealing, normalizing or stress relieving.

Steel containing up to 0·83% carbon

The slow cooling of a piece of steel of this composition gives a final structure of ferrite and pearlite. The transformation from austenite to ferrite and pearlite takes place over a definite temperature range. For the transformation to take place correctly, it is important for the cooling rate to be extremely slow.

The extreme opposite of slow cooling would be rapid cooling brought about by quenching the material in water. If the steel is above the upper transformation temperature when quenched, the carbon does not have sufficient time to take up its correct position in the material, therefore the pearlite will not form. Also, on rapid cooling the structural change from face centred cubic to body centred cubic atomic arrangement takes place at a lower temperature—between 350°C and 150°C. This means that the carbon is squeezed and trapped in the body centred cubic atomic formation of the iron. The atomic structure of the iron is distorted and there is little space between the atoms. The physical effect of this is hardness and brittleness of the material. When viewed with a microscope, the final structure is not of ferrite and pearlite, but martensite.

Steel containing 0·83% carbon

When slow cooled, a steel of this composition forms a structure which is entirely pearlite. The change from austenite to pearlite takes place at one temperature.

If the material is quenched from above this temperature, pearlite has insufficient time to form, and the resulting structure will be martensite.

Steel containing above 0·83% carbon

With slow cooling, the structure formed is of pearlite and cementite. If the material is quenched from above the upper transformation temperature then the structure will be martensite.

All plain carbon steels containing in excess of 0·3% carbon

Heat-treatable carbon steels are softest when they are slow

cooled through the transformation range. When the cooling rate is increased then the steel becomes harder.

When welding is carried out on hardenable carbon steels there are several problems which may arise, but can be overcome by correct heat treatment. This heat treatment may be in the form of preheating or postheating.

Preheating

Preheating is carried out for various reasons:
(1) The heat losses are reduced from the weld area, which in turn reduces the cooling rate of the weld.
(2) If preheating is not carried out, there is a drastic fall in temperature between the weld area and the parent plate. This may cause rapid cooling, leading to the formation of martensite and probable cracking.
(3) The expansion and contraction rates are reduced, therefore there is less distortion and residual stress.
(4) With increased temperature the conduction rate is reduced.

Postheating

After welding, some form of heat treatment may be necessary. Depending upon the requirements of the finished article, the following treatment after welding may be carried out: annealing, normalizing, hardening, tempering, stress relieving or spheroidizing.

Annealing

During welding, the size and the conditions of the grains of which the material is composed may be seriously affected. Depending upon the welding process used, the grains of the material may be large in size, or they may be distorted due to the stresses set up during and after welding. The stresses are corrected by annealing and the grain structure refined, so that the material becomes softer and free from residual stresses, enabling it to be worked or machined.

When steel is annealed, it is heated to 40°C above the upper transformation temperature, which varies with the carbon content of the steel. It is held at this temperature long enough for the carbon to distribute itself evenly throughout the austenite. For most practical purposes it is held at the annealing temperature for one hour per one inch thickness of material. The steel should be cooled slowly, preferably in the furnace or buried in hot ashes or lime. After annealing, the structure of steel with a carbon content of less than 0·83% should be grains of pearlite with grains of ferrite.

Normalizing

Very often the terms normalizing and annealing are misquoted. When normalizing, the steel is heated 40°C above the upper transformation temperature for the particular composition of the material being normalized. It is only held at this temperature a sufficient time for the material to transform to austenite. The steel is then allowed to cool in still air. Normalizing is faster than annealing and is often used in the welding industry to refine any coarse grain structure, to reduce stresses after welding or to remove any hard zones in the heat-affected area. The grain size of a normalized structure is smaller than that of annealed steel, giving a normalized structure increased toughness. A typical example of normalizing being carried out after welding is in the case of mild steel pipes. The pipes are heated to between 900° and 950°C, this temperature being held for a period of two minutes for pipes up to 4 in diameter bore, and for five minutes with over 4 in diameter bore. This is followed by slow cooling in still air.

Quench hardening

In quench hardening the steel should be heated 40°C above the upper transformation range and held at this temperature long enough for the internal structure to form austenite. Overheating and prolonged heating should be avoided in order to prevent grain growth. The increased hardness may be obtained by quickly cooling the steel by immersing it in oil, water or another quenching medium. When quenching is being carried out, the article to be hardened should be agitated in the cooling liquid. This is to prevent gases or steam forming around the article and interfering with the quench rate. The successful quench will convert the austenite to martensite.

Tempering

A quench hardened steel with a structure of martensite is very brittle. If it was in the form of a chisel and was struck with a hammer, it would fracture in a brittle manner. To relieve this brittleness, the article should be tempered. This is done by reheating the hardened material, usually to a temperature between 200° and 450°C, generally followed by quenching. The effect of tempering is to obtain a tougher material, hence there is a decrease in the brittleness and a slight decrease in hardness.

Stress relieving

Stress relieving may be carried out after welding or excessive cold working. The purpose of stress relieving is to remove any internal or residual stresses which may be present in the material. These stresses can be relieved by uniformly heating the welded component to a temperature which is below the transformation range. The welded steel components are heated to a temperature of between 600° and 650°C, and they are usually held at this temperature for one hour per one inch of thickness, followed by slow cooling. If an examination of the microstructure is carried out, it can be seen that recrystallization of the ferrite has taken place, which contributes to the relaxation of the residual stresses of the welded joint.

Another form of stress relief may be carried out by heating the component to about 450°C for a period of time. At this lower temperature very little, if any, recrystallization of the ferrite will occur. If the welded component is held at 450°C for one hour, and slowly cooled, about 40% of the stresses will be relieved. This may be regarded as adequate for some purposes.

It is of interest to note that any gases trapped in the cooling welds may be released from the welded components during the stress relief heat treatment.

Spheroidizing

This renders high carbon steels softer and more easily machinable than does annealing. The steel is heated to just below the transformation range and held at this temperature for a number of hours. Cooling should be allowed to take place very slowly throughout the upper part of the cooling range. The cementite collects into spherical-shaped particles, leaving ferrite throughout the remaining structure. After machining, the steel is reheated to put the carbides back into solution; this means heating to the austenite condition. Normal heat treatment can then be carried out.

Case hardening

In order to increase the wear resistance of a piece of mild steel, its surface may be treated to give increased hardness whilst the interior remains in the soft condition. This may be done by:

(*a*) Case carburizing. (Pack hardening and cyanide hardening.)
(*b*) Nitride hardening.

Pack hardening
This process involves the introduction of carbon into the surface of a piece of mild steel. The parts to be surface hardened are packed into a container along with a material which contains a high content of carbon. Parts to be hardened should be perfectly clean, whilst parts which do not require hardening should be protected from the carbon-containing materials. This may be done by copper plating, or coating with a material which will prevent carbon from entering the surface of the component. The boxes or containers are then sealed and placed in a furnace, the temperature of which is raised to about 920°C. At this temperature the steel is in the austenitic condition, with its structure face centred cubic, which is capable of absorbing carbon whilst in the solid state. The components are maintained at this temperature for a number of hours. During this time the carbon migrates from the material in which the parts are packed into the surface of the parts to be hardened. This is followed by slow cooling in the furnace. Upon removing from the furnace, the steel will possess the following characteristics:

(1) A surface case of high carbon steel will contain about 0·83% carbon, the depth of this being up to $\frac{1}{16}$ in.
(2) The core will contain its original carbon content.
(3) The entire structure will have large grains through being maintained at a temperature above the upper transformation range for too long a period of time.

It is now apparent that the steel has two separate percentages of carbon and therefore must have two A.C.3. points. Heat treatment

may now follow for the following reasons:
(1) To refine the grain size, thus increasing the toughness of the core.
(2) To produce a case of hard wear resisting martensite.

To refine the grain size of the core, it is necessary to reheat the components to above the transformation point of the inner core, i.e., to about 880°C. This temperature is held for a sufficient period of time for the grain refinement of the core to be completed. This will then be followed by quenching in water or oil. Despite the small grain size of the core, which is desirable, the structure of the core will be of ferrite and very small amounts of martensite, which is undesirable. Furthermore, the case, owing to it being reheated well above its own upper transformation point, is now martensitic and coarse. Therefore, a second heat treatment cycle is required. This consists of reheating to just above the transformation point of the case, i.e., 750°C. The grain structure of the case will be refined as the newly formed grains of austenite are produced. By quenching from this temperature, a fine-grained martensitic case is obtained which is very hard but not liable to chipping because of excessive brittleness.

The effect of this second heating and quenching is most beneficial to the core of the component. The martensite is reduced and a tougher and more ductile structure replaces it.

Cyanide hardening

This process is faster than pack hardening and it offers the following advantages:
(1) It is a much cleaner method of case hardening.
(2) Parts to be hardened need not be fully immersed in the liquid salt bath, therefore parts which do not require hardening can be left unprotected outside the hardening liquid.
(3) There is less distortion compared with pack hardening.
(4) Because the heated parts are immersed, scaling on the surface is unlikely to occur during the hardening treatment.
(5) With this method of case hardening, it is possible to immerse each item separately, this gives distinct economical advantages when there are only a few components to be surface hardened.

The parts to be case hardened are cleaned and dried and then suspended in a liquid salt bath containing a large percentage of potassium or sodium cyanide at a temperature of 900° to 1 000°C.

The immersion time can last from about half an hour to two hours, depending upon the depth of the case required, which may be from 0·004 to 0·01 in. On removal from the salt bath the components may be quenched in water, which has the two-fold effect of producing the desired hardness and cleaning the surface of the component.

When using this form of case hardening, it should be realized that any form of cyanide is a deadly poison and gives off toxic fumes; the necessary precautions must be adhered to for maximum safety.

Nitride hardening
Iron combines with nitrogen forming a hard, brittle compound known as iron nitride. This compound can be used to advantage when surface hardening certain alloy steels. These alloy steels are specially manufactured to allow controlled hardening to take place by nitriding.

When nitriding, the parts to be hardened are placed in a gas-tight container and heated to a temperature of 500°C. This temperature is maintained for a period of from 48 hours' to 100 hours' duration. Whilst at this temperature, ammonia gas is fed into the container. Ammonia is a compound of nitrogen and hydrogen, and the effect of the temperature allows the ammonia gas to be broken down into its atomic state as nitrogen and hydrogen. Nitrogen is the smallest of all atoms and readily enters the solid steel at the seemingly low temperature. The nitrogen atoms which enter react with the iron to form iron nitride, and, to a lesser degree, nitrides with other elements which are present in the steel. A hard surface case is therefore formed to a depth of approximately 1 mm. The advantages of nitride hardening include the following:

(1) It is retained at a higher temperature than a case which has been hardened by carburizing. A pack hardened case would begin to soften at 150°C whilst a nitride case remains hard up to 500°C.
(2) Heat treatment is not required because the component is not heated to the transformation range, so that there is no alteration of the grain size and structure.
(3) The presence of the alloying elements in the steel and the hardened case offers advantages in the form of resistance to corrosion and fatigue failure.
(4) The low temperature at which nitriding is carried out ensures that there will be little or no distortion of the components whilst case hardening.

Welding Science and Metallurgy

Unfamiliar terms used in this chapter

Hardenable: Capable of being hardened by heat treatment.
Transformation: To change the internal structure when heating or cooling a piece of steel.
Residual stresses: Stresses remaining in a structure as a result of heat or mechanical treatment.
Toxic: Poisonous.

Chapter review questions

State the advantages of nitride hardening over pack hardening.
What is the difference between annealing and normalizing?
Why does a pack hardened steel have two A.C.3. points?
How may a piece of steel be hardened when subjected to heat treatment?
If a piece of 0·83% carbon steel is quenched from 750°C, why will the impact resistance be low?
How deep is the case when (a) pack hardening, (b) nitride hardening?
What heat treatment process will soften a piece of steel?
List the benefits that may be gained by heat treatment.

CHAPTER 4

Consideration of Effects during the Welding of Iron and Steel

Fusion weld in iron

When fusion welding is carried out, the plates being welded are melted at their fusion faces and liquid filler metal is added. The source of heat may be an electric-arc or oxy-acetylene flame. It is important that the weld metal should be protected from the atmospheric gases, oxygen and nitrogen. Whatever modern form of welding is used, a gaseous shield is provided around the arc to prevent atmospheric contamination, in the form of:

(1) A gaseous shield produced from a flux coating which is around the electrode core wire. A typical example of this is when welding is being carried out by the manual metal-arc process.

(2) An artificial atmosphere can be created around the arc to provide protection, possibly in the form of carbon dioxide supplied from a cylinder.

(3) Several automatic electric-arc welding processes use granular flux to protect the weld metal. The heat of the arc changes the granular flux to the gaseous state providing a suitable shield.

(4) When oxy-acetylene welding, the reducing gases, carbon monoxide and hydrogen which are produced in the flame, protect the weld pool from the atmosphere, providing that the flame is neutral.

The solidification of the weld metal may now be considered without interference from the atmospheric gases. When considering the weld pool, it will always be at a temperature above the melting point of the iron. At the same time the liquid iron is in contact with solid iron which is at a temperature of 1 530°C, the freezing point of iron. (See Fig. 17.)

Welding Science and Metallurgy

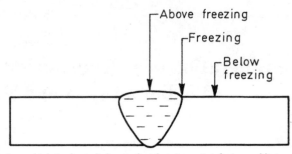

Fig. 17. Iron which is above, below and at freezing point during welding.

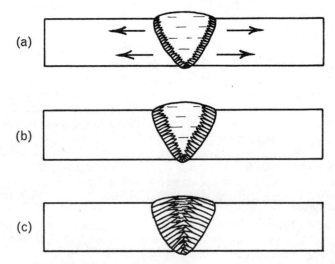

(a)

(b)

(c)

Fig. 18. (a) Commencement of freezing. Arrows show heat lost by conduction.
        (b) A further stage in freezing.
        (c) Completion of freezing.

It is now time to consider the ways in which heat may be lost from the weld area. A small amount of heat is lost to the surrounding atmosphere. This is negligible, the major portion of the heat being lost by conduction through the mass of the iron being welded. When welding is carried out using the manual metal-arc process, the arc moves along the joint away from the metal just deposited. Because the source of heat moves away and heat is conducted away, the weld pool will cool and solidify. Fig. 18(a) is typical of a single run weld on iron.

32

It will be seen that the solidification has commenced along the line which is at the freezing temperature of the iron. Because the heat input by this process is localized and the plates have not been preheated, cooling will take place and cause the grains to form in the direction opposite to the heat losses. This means that the liquid atoms of iron are continuously being cooled and they solidify by attaching themselves to the grains of the iron in a solid state. (See Fig. 18(*b*).) This action continues, the liquid atoms now attach themselves to the grains of the weld metal which has just solidified. This means that the grains are growing inwards and upwards at the expense of the liquid metal. (See Fig. 18(*c*).) The grains formed are distinctly columnar and each continues to grow until its growth is restricted by contact with another. Heat travels through liquids by convection currents, and liquid iron is no exception. This explains why the last of the liquid iron solidifies at the top of the weld to complete the formation of columnar grains.

Welding heat and its effects on the iron base plates

As the temperature in the weld metal falls, the temperature of the plates being welded increases. This results in a steady temperature gradient from the weld. If temperature readings are taken at regular distances from the weld, the temperature is found to be decreasing from 1 530°C to room temperature. All the iron which is heated above 910°C will undergo an internal atomic structural rearrangement, from body centred cubic to face centred cubic. If Fig. 19 is considered, all the metal between 1 530°C and 910°C will have undergone the internal change. The grains nearest to the weld will have been heated to the highest temperature and, unfortunately, will have increased in size. All the iron that does not reach the

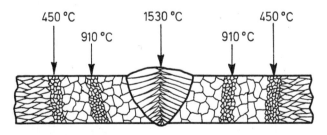

Fig. 19. Effects of welding heat on iron. Microstructure × 100.

temperature of 910°C will not have undergone any atomic rearrangement, but, if the iron has previously been cold worked, then the distorted grains will have recrystallized at a temperature above 450°C.

Consider the macrostructure of a single run of weld deposited between two pieces of iron which is in the cold-rolled condition. There are three distinct zones, as seen in Fig. 20.

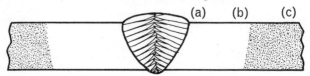

Fig. 20. Effects of welding heat on iron. Macrostructure × 4.

(*a*) The columnar grains of the weld metal, which can be clearly seen.

(*b*) The area next to the weld, which is known as the heat-affected zone.

(*c*) The area unaffected by the heat of the arc, which will still be in a cold-rolled condition.

It is now necessary to consider the microstructure of the heat-affected zone in order to see what has taken place. Because the high temperature of the arc raises the temperature of the iron to well above 910°C, the grains will merge together near the weld. This grain growth is seen to decrease gradually in size as the distance from the weld increases. Another change, due to the heat, is the fact that recrystallization of the cold-rolled plate takes place in the areas which exceed 450°C during welding. It will be observed that the iron which is not heated to above 450°C is still in the cold-rolled condition.

Multi-run welds in iron

If a weld is deposited into a single vee butt joint, for example, between two pieces of ⅜ in thick plate, the first run of weld to obtain penetration is of a columnar grain structure. (See Fig. 21(*a*).) After chipping and brushing, the iron will have cooled considerably. To obtain the correct build up and reinforcement of weld metal, a second run of weld is then deposited on top of the first run. The heat from the second run of weld causes the metal deposited in the first run of weld to be heated above 910°C. When heated above this temperature the columnar grains in the first run recrystallize to form small equiaxed grains. On cooling, the newly formed grains remain in the equiaxed condition. However, the second run of weld

will be composed of coarse columnar grains and can only recrystallize to equiaxed grains by depositing another weld bead or
by subsequent heat treatment, i.e., normalizing. Fig. 21(*b*) shows
that the first run of weld has reached 910°C, the transformation
temperature, and has formed new equiaxed grains. In Fig. 21(*c*)
the entire weld structure has been heated to just above 910°C
causing the transformation to occur throughout, the entire structure being formed of fine equiaxed grains. This is a typical example
of normalizing having been carried out.

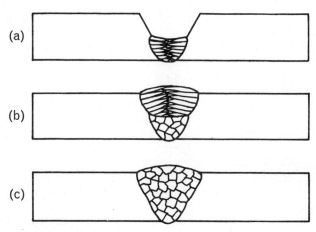

Fig. 21. (a) *First run of weld, showing columnar grains.*
(b) *Transformation of the first run of weld.*
(c) *Heat treatment by normalizing, giving equiaxed grains throughout.*

Mild steel

Mild steel contains up to 0·3% carbon in addition to iron. The
presence of carbon makes a considerable difference to the sequence
of events which takes place on heating and cooling. In considering
a piece of welded mild steel, welded by a fusion welding process, the
material will have been melted in the weld itself, but the heat will
have been lost from the weld into the mild steel plates which have
been welded. In turn, this means that the temperature of the mild
steel plates has been raised to a range of temperatures, decreasing
from the actual weld. The manner in which the heat is lost is
similar to that lost from the weld with iron. The welded mild steel

35

plates therefore show a heat-affected zone, the extent of this depending upon several factors, the first of which is the source of heat for welding. The heat source may be conveniently grouped into two kinds:

(1) The oxy-acetylene flame with its heat intensity of 3 200°C, the quantity of heat being decided by the choice of nozzle. In bringing the mild steel plates to melting point, quite a large area around the weld pool will be at a high temperature, the reason for this being that it takes time to bring the mild steel to melting point.

(2) The electric arc with its heat intensity of 4 000°C, the quantity of heat being dependant upon the amperage and voltage. The heat source in this case is local and intense, and the heat transfer into the plates is very efficient. When manual metal-arc welding, the weld pool is formed instantaneously and the speeds of welding are much faster.

If the two heat sources are compared, the rate of welding is much slower by oxy-acetylene welding, and plenty of heat is lost by conduction from the weld into the plates of mild steel. The effect of this is to cause the weld metal and the base metal to cool at a slow rate. On the other hand, the cooling rate of a manual metal-arc weld is quite rapid. It must be understood that the cooling rates of a weld directly influence the structure of the weld and heat-affected zone, and the mechanical properties.

From the above, it will be appreciated that it is not possible to decide upon one rate of cooling or to state that all mild steel welds cool at a certain rate. But it is possible to state that, regardless of how far the heat-affected zone extends, some of the mild steel will be heated to a temperature well above the transformation range, some just above, some in between the transformation range and some below it. It is very probable that the mild steel plate will have been cold rolled, and even the heat which was below that of the transformation range will have an effect on the structure and the mechanical properties. With this in mind, it is now possible to give consideration to welds and base material made with slow or fast cooling conditions. It is also important to note that any type of weld and its heat-affected zone can be altered if there is subsequent heat treatment.

Slow cooling

When carrying out a weld by the oxy-acetylene process on two pieces of cold-rolled mild steel, each piece being $\frac{3}{16}$ in thick, heat is applied to the plates at the intended start of the weld. Sufficient heat is applied to bring the edges of the plates to the molten state and to overcome the heat which is lost by conduction. The temperature of the weld pool will be over 1 450°C. If it were possible to take a micro photograph of the condition of the mild steel whilst welding is in progress, it would be seen that there are various structures in the mild steel. These structures are created by the temperature gradient, which is decreasing from the weld pool.

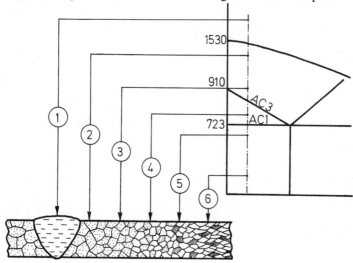

Fig. 22. Structures observed when a cold-rolled mild steel plate is in the process of being welded.

Consideration should now be given to Fig. 22. The iron-carbon equilibrium diagram is used to illustrate what is actually taking place at the various temperatures to which the plates have been heated during the course of welding.

Arrow 1. This indicates the molten weld pool, the exact temperature of which cannot be predicted accurately. But it is certain that it will be above 1 450°C, which is the melting point of the mild steel. The temperature of the solid material at the edge of the weld pool will be at the freezing temperature of the mild steel.

Arrow 2. Here the structure is austenite. The mild steel has been heated to well above the transformation range, which causes the grains to merge one into another, producing large coarse grains.

37

Arrow 3. The structure of the mild steel is still austenite, but it will be seen that the grain size is smaller. The reason for this is that the material has only been heated to a temperature just above the transformation range. Because of the lower temperature to which the material is heated, compared with Arrow 2, there is less chance of grain growth taking place.

Arrow 4. In this case the mild steel is heated to a temperature within the transformation range. At this temperature the pearlite will have changed to austenite, this having occured at 723°C. The remaining ferrite is in the process of being absorbed by the austenite.

Arrow 5. This shows the structure of the mild steel which has been heated to a temperature of 600°C, i.e., below the transformation range. Here the structure is composed of ferrite and pearlite. Because the temperature of 450°C has been exceeded, the cold-rolled distorted grain structure will have recrystallized, forming new equiaxed grains of ferrite. The harder pearlite structure remains unchanged because it has not been heated to A.C.1. temperature.

Arrow 6. At this temperature, below 450°C, the cold-rolled mild steel structure remains unaltered. The reason for this is that the recrystallization temperature has not been reached.

It is now possible to consider what will occur upon slow cooling the mild steel to room temperature. When the blowpipe passes along the weld seam, solidification of the weld metal occurs. The reducing gases of the flame still protect the weld area, ensuring that it is not exposed to the atmosphere whilst solidification takes place. When the liquid weld metal is in the process of cooling, small columnar crystals form where the liquid mild steel is in contact with the solid mild steel. These grow inwards and upwards into the liquid weld pool. At the same time dendrites appear in the molten weld pool. (See Fig. 23.)

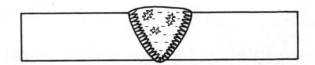

Fig. 23. Commencement of solidification.

The columnar crystals continue to grow in size towards the centre of the weld pool, whilst, at the same time, the growth of the dendrites continue. The temperature of the weld pool remains constant until solidification is complete. The resulting final structure is composed of coarse columnar crystals and equiaxed crystals, as shown in Fig. 24. The reason for the presence of any equiaxed crystals is that they have time to form in the liquid weld pool because of the slow rate of solidification. The term 'slow rate' is intended to mean that the solidification time of a weld made with the oxy-acetylene process is longer than the solidification of a similar weld made with the manual metal-arc process.

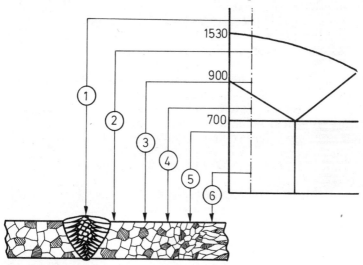

Fig. 24. *Solidified weld at room temperature.*

The structures at the various temperature zones when the weld metal was liquid, as shown in Fig. 22, can now be examined when cooled down to room temperature. Referring to Fig. 24, the temperature zones have been given the same arrow number for the purpose of comparison. One might expect austenite to follow the normal transformation pattern on cooling, i.e., the precipitation of ferrite starts to take place at the A.R.3. temperature and continues until the A.R.1. is reached. At this temperature the eutectoid reaction takes place, and the austenite breaks down to ferrite and cementite, this being known as pearlite. Therefore the final structure of the mild steel is ferrite and pearlite. The correct formation of the ferrite and pearlite is responsible for the characteristic toughness of the material. A normalizing treatment, for example, provides the correct toughness, the resulting structure being pearlite evenly distributed in the ferrite. It must be remembered, however, that a single run of oxy-acetylene weld is not a normalizing treatment, mainly because of heating to too high a temperature.

Consideration may now be given to Fig. 24. The various temperature zones, noted when heated during welding, may now be seen individually at room temperature.

Arrow 6. The microstructure indicates that this area has remained unaffected by the heat of welding, the maximum temperature reached in this area being 250°C. When fully cooled, the increases in hardness and tensile strength which were brought about by cold rolling still remain in the mild steel plates.

Arrow 5. The mild steel in this region has been heated to about 600°C by the welding heat. Because there has been cold rolling of the mild steel prior to welding, recrystallization of the ferrite occurs. It should be noticed that the final structure, when cold, consists of newly formed fine equiaxed grains of ferrite, along with pearlite which remains unaffected at this temperature. The ferrite which recrystallized is a contributing factor to the relief of residual stresses.

The temperature of 600°C is also considered a very suitable temperature for stress relieving of welded joints. The internal stresses caused by cold working, and stresses caused through welding can be relieved at this temperature.

Arrow 4. The structure to be seen in this region is of ferrite, in two distinctly different grain sizes, and pearlite. During welding, the mild steel is heated to a temperature within the transformation range, in this case 770°C. When at this temperature, the pearlite changes to austenite, and the absorption of the ferrite by the austenite is in progress. This absorption of the ferrite does not go on to completion, forming an entirely austenitic structure, because the temperature is not high enough for this to occur. When cooling takes place, the ferrite which is not absorbed by the austenite remains unchanged and may be identified as the largest ferrite grains in the microstructure. The smaller grains of ferrite are produced as follows. When cooling takes place, the austenite begins to precipitate out the ferrite which was absorbed during heating. This occurs within the boundary of the austenite, and continues until the A.R.1. is reached, when, at this temperature, the eutectoid reaction occurs and the remaining austenite transforms to pearlite. This explains why the final structure consists of ferrite, with its two distinct grain sizes, and pearlite. This type of structure is quite acceptable.

Arrow 3. The heating during welding of this region was sufficient to form austenite, the mild steel being heated a few degrees above the transformation range. On cooling, the events which take place are very similar to normalizing. Namely the austenite precipitates the ferrite out on slow cooling until the A.R.1. is reached, and the eutectoid reaction then takes place, leaving a final structure at room temperature of ferrite and correctly formed pearlite.

Arrow 2. The origin of the microstructure shown in this region is an austentic structure produced by receiving intense heat from the weld pool. Unlike the grains of austenite shown under Arrow 3, Fig. 22, these are quite large. (See Arrow 2, Fig. 24). On cooling, a large grained austenitic structure will cool and ferrite will be precipitated out at the grain boundaries and in the grains of the austenite. This takes place whilst the mild steel is at a temperature within the transformation range. When the A.R.1. is reached the pearlite forms, but it is split up by the ferrite, as may be seen from an enlarged microstructure as shown in Fig. 25(c). Widmanstätten is the name given to this type of structure. It will be explained along with the structure noted on cooling under Arrow 1.

Arrow 1 and Arrow 2. The weld pool has solidified and it has a coarse grained and overheated structure. Examination of the grains shows an 'as cast' structure. Overheating is caused by heating the material to too high a temperature above the A.C.3. temperature for the carbon composition of the steel. The result of this is grain growth and the formation of large austenite grains during welding.

The microstructure in Fig. 24 (Arrow 1) only shows the grain boundaries. This is deliberate for the structure formed cannot be detailed at this scale. A separate detailed view of this will be considered. Fig. 25(a) shows that at a temperature well above the A.C.3. the grains are quite large, and they are in the austenitic condition. On cooling to the A.R.3. temperature, ferrite starts to precipitate from the austenite. Owing to the grains being quite large, the ferrite deposited is at the grain boundaries; this is because the ferrite cannot group together to form its own separate grains. The reason for this is that the distance from the middle of the grains to the boundaries is too far. When the grain boundaries are completely surrounded by the ferrite, any of the ferrite remaining in the grains of austenite is precipitated out within the grain itself. (See Fig. 25(*b*).) This occurs within the transformation range. Reference may now be made to the structure after the eutectoid reaction as taken place, i.e., below the A.R.1. temperature. (See Fig. 25(*c*).) It can be seen that the pearlite is scattered about, being separated by the ferrite. The effect of this is to reduce the toughness of the steel. This coarse, undesirable structure is known as a Widmanstätten structure.

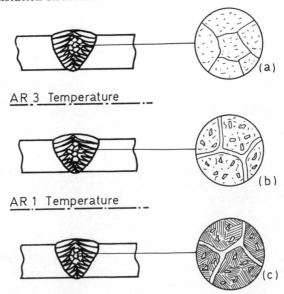

AR 3 Temperature

AR 1 Temperature

Fig. 25. The stages of cooling of an overheated weld structure.

The undesirable structures left in a weld may be corrected by heat treatment. Often this is carried out in the welding workshop by reheating the weld to red hot and carefully hammering it. The grain structure is broken down and, since it is well above its recrystallization temperature, small equiaxed crystals form, with good mechanical properties. Small diameter, round bars are often welded by the oxy-acetylene process, and are subsequently hammered, and are then allowed to cool slowly. Another form of heat treatment is normalizing or annealing, this producing a fine-grained equiaxed structure. It is common practice, when welding has been carried out on high pressure steam pipes by the oxy-acetylene process, for a normalizing treatment be used. The blowpipe flame reheats both the weld and the heat-affected zone. Slow cooling, away from any draughts, is essential for producing a fine-grained tough structure.

Multi-run welds by oxy-acetylene

No mention has yet been made of a second or subsequent layer of mild steel weld metal being deposited. When multi-run or block welding is carried out on thick sections of mild steel, more than one layer of weld metal is necessary. The first run of weld is of a coarse 'as cast' structure. When the second run of weld is deposited, the heat from the blowpipe flame raises the temperature of the first layer through the transformation range, causing recrystallization and grain refinement. However, there is an increase in the width of the heat-affected zone, and this in turn causes a slower rate of cooling of the weld metal. Because of this slower rate of cooling, the refined grains in the lower run of weld tend to grow a little in size before the weld has completely cooled. In consequence, it must be regarded as an acceptable structure but, at the same time, it should be realized that the grain size will be larger than a similar first run which has been reheated by manual metal-arc welding. It can now be seen that the first or lower runs of weld metal undergo grain refinement, and provided that they are heated through the transformation range, the resultant toughness of the metal in the lower runs of weld increases. The final layer of weld is a coarse grained 'as cast' structure, which can only be improved by correct heat treatment.

Fast cooling

When carrying out a weld with the manual metal-arc welding process, the cooling rate of the deposited weld metal is much faster than when using the oxy-acetylene process. Another major difference between the two processes is the range of thickness of mild steel which may be welded. Multi-run welds are rarely carried out by the oxy-acetylene process, whilst the manual metal-arc process is often used.

Consideration may now be given to a weld being carried out between two pieces of ⅜ in thickness mild steel plates, using the manual metal-arc process. Fig. 26(a) shows the penetration run of weld deposited between the mild steel plates, these being in the cold-rolled condition before welding. The weld metal deposited is composed of columnar crystals, the rate of cooling being rapid. Whilst it is difficult to assess a definite cooling rate for a first run of weld deposited by the manual metal-arc process, it can be regarded as being just slower than an oil quench from a high temperature. The coarse columnar grains do not give the most desirable structure obtainable by manual metal-arc welding. The

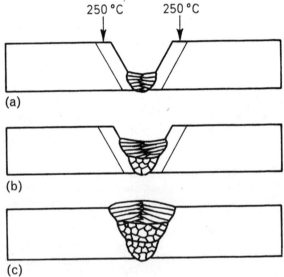

Fig. 26. *Runs of weld deposited, and their effect.*

effect of the heat lost into the plates from the first run of weld is to preheat a small region at each side of the weld. This heat helps to reduce heat losses from the weld metal when subsequent runs of weld metal are deposited. Any undesirable effects in the first layer of weld metal are removed by the reheating caused by heat from subsequent runs of weld being deposited. This is shown in Fig. 26(b) where it is evident that a refinement of the first run of weld has taken place. This is very desirable because the toughness of the weld metal has been improved by being reheated through the transformation range, and by changing its internal structure to fine equiaxed grains. From Fig. 26(*b*) it is evident that the second run of weld deposited is made of coarse columnar grains of 'as cast' weld metal. This second run of weld puts further heat into the base plates being welded.

If a final run of weld is placed over the second deposit, then the heat refining cycle will again take place. The last run of weld to be deposited will always be of coarse columnar crystals, which can only be refined by normalizing or annealing. (See Fig. 26(*c*).)

The heat-affected zone is a little wider as a result of the third and final run of weld. Consider a welder depositing runs of weld in the order just described. Immediately after the first run of weld is laid, he deslags and wire brushes the deposited metal. The layer of slag which is on the surface of the weld will tend to prevent heat losses to the atmosphere. By the time the second run of weld is about to be made, the temperature of the mild steel plates in the heat-affected zone will have cooled to around 250°C. The next layer is then deposited; this is a little heavier than the first one and, in consequence, more heat is added to the plates. The same would occur with a third run of weld.

The completed weld consists of three layers of weld metal, two of which consist of fine grained weld metal, and the third of which is a coarse grained final layer. The heat-affected zone and the weld can now be considered. The rapid and efficient way of heating the mild steel plates to melting point when manual metal-arc welding results in heat being lost to the plates by conduction. The heat from the weld is rapid and a heat-affected zone can be seen. By comparison with the heat-affected zone which was shown after oxy-acetylene welding (see Fig. 24), the manual metal-arc heat-affected zone is quite narrow, and rarely exceeds one centimetre on either side of the weld. Some of the arrows, used to indicate the changes in microstructure when oxy-acetylene welding, with its slow cooling, are quite close to each other when considering manual metal-arc welding. These are Arrows 2, 3, 4 and 5 on Fig. 24 and Fig. 27.

When considering Fig. 27, it should be assumed that a correctly flux-coated electrode has been used, along with correct current and voltage. The cold-rolled mild steel plate and weld metal have again been split up into six zones, each being indicated by a numbered arrow. Each will be considered individually.

Arrow 6. This simply indicates that the cold-rolled plate was unaffected by the heat from welding.

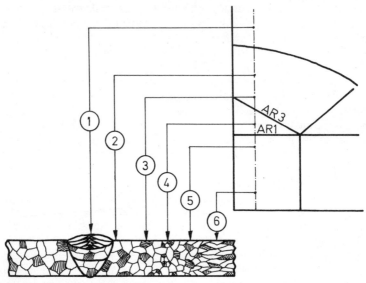

Fig. 27. Changes in microstructure after manual metal-arc welding.

Arrow 5. The mild steel has been heated to a temperature in excess of 450°C during welding. At this temperature range, recrystallization of the ferrite occurs. When cooled, this gives a final structure of fine equiaxed grains of ferrite, along with pearlite.

Arrow 4. During welding, the steel is heated to within the transformation range. On cooling, any austenite, which forms when heating, reverts back to ferrite and pearlite, giving a final structure of ferrite in two distinctly different grain sizes, and pearlite. For a description of how this occurred, see p. 40.

At this stage, precisely the same microstructures have occurred as when oxy-acetylene welding was considered along with its slow cooling rates. But when the manual metal-arc process is considered, the arrows of the different heat-affected zones are much closer to the weld, thus indicating a smaller heat-affected zone.

Arrows 3 and 2. Here, the steel, shown by Arrow 2, which is immediately adjacent to the fusion zone, is subjected to a temperature which normally encourages the grains to grow, developing into a coarse structure. (See Fig. 24 Arrow 2.) It can be seen that this is a structure of a few large grains, instead of the numerous small grains which are desirable. However, when manual metal-arc welding, little difference in grain size is observed between the microstructure in Fig. 27, Arrow 2, and the microstructure of the steel which had been heated to only a few degrees above the transformation range while welding was being carried out in Arrow 3 Fig. 27. The reason for this is that grain growth is at a minimum with this process. During welding, the steel was rapidly heated, followed by rapid cooling, and was thus being kept at a high temperature for a very short period of time, so keeping down the size of the grain. This very narrow zone does not show any sign of the mild steel having been cold worked; the reason for this is that the temperature of the arc is sufficient to raise the steel through the transformation range which results in a refinement of the grain structure.

Arrow 1. The weld zone: when the cycle of heat treatment has refined the grain structure, there is an improvement in the toughness of the mild steel. The weld shows two layers of equiaxed grains and a final layer of the columnar type of grains.

From the microstructure, the chemical composition of the weld metal can be somewhat deceiving. The carbon content of the weld metal will be less than the cold-rolled mild steel plates, although this is not apparent in the microstructure. This loss of carbon occurs when in transit across the arc on the way to the weld pool. Therefore, other ways of increasing the hardness and strength must be used to compensate for the carbon losses. One of these methods is to use alloying elements which are included in the electrode which transfers across the arc—small percentages of manganese are often used. A small amount of nitrogen, not exceeding 0·02 to 0·03 %, may enter the weld where it will form iron iron-nitride eutectoid. This resembles pearlite when viewed with the microscope, and is distributed throughout the structure, giving the desirable mechanical properties. It is only when manganese is present that the iron iron-nitride eutectoid forms. If manganese is not present, then undesirable iron nitride needles form. The widespread opinion that nitrogen should be kept out of mild steel welds altogether is not quite true. When a small controlled amount, such as 0·02 %, is allowed into the weld, beneficial results are obtained. This must not be confused with the excessive amount of nitrogen which may enter a weld when welding with a bare wire or a conventional electrode with a broken flux coating. Electrodes are manufactured to allow the required amount of nitrogen to enter the weld.

Summary

Mild steel, for all practical purposes, has good weldability. It can be readily welded by all known welding processes. Before welding, it is necessary to give consideration to the protection of the weld from the atmosphere. The quantity and intensity of heat of the welding process being used has an effect on the structure of the material being welded. The length of time at high temperatures and the resulting cooling rate alters the size of the heat-affected zone, and this in turn interferes with the desired mechanical properties. On a single run weld made by the oxy-acetylene process, subsequent heat treatment may be necessary in the form of normalizing. When normalizing is carried out, the heat-affected zone and the weld metal should be heat treated. When manual metal-arc welding sections which require more than one run of weld, to obtain the required height of build up, the heat treatment cycle is virtually carried out by the deposition of subsequent layers of weld metal.

A job left to cool, when partially welded, is an example of how the very desirable grain refinement, when arc welding thick sections, may be lost. By recommencing welding on cold steel plates, subsequent layers only refine the immediate top surface of the previously deposited weld metal. The advantage of heat being put in the steel by the previous welds is lost. With an understanding of the information that is revealed in this chapter, the undesirable structures which can be produced when welding may be controlled adequately.

Unfamiliar terms used in this chapter

Contamination: To introduce impurities.
Toughness: Combined strength and ductility.
Recrystallization: The heating of a cold-worked material to a temperature which is high enough to allow a new grain structure to form.
Eutectoid: A solid solution decomposing to a mixture.
Overheating: Heating to a temperature above the transformation range for too long a period of time.
Grain refinement: To change the grains to a small equiaxed condition.
Iron nitride: A hard brittle compound.

Chapter review questions

What type of gaseous shields are used to protect welds from the atmosphere?

Which way is heat likely to be lost from a weld?

What is the approximate recrystallization temperature of ferrite?

When oxy-acetylene welds are slow cooled, what type of grains will form?

What would be the effect of block welding on the structure?

Why is the heat-affected zone less on an arc weld than on an oxy-acetylene weld?

What will be the effect of depositing multi-runs when arc welding?

In which way does an iron-nitride eutectoid improve an arc weld?

CHAPTER 5

Weaknesses when Fusion Welding Mild Steel and Hardenable Steels

Mild steel has earned its reputation as being the best material for fabrication work. Its strength and hardness make it ideally suited for such work, whilst its ductility is such that welding may be readily carried out. 'Welds which are made in mild steel are easily made': this, or words to this effect, is a common statement. In reality, these welds require the most thought and care. When welding is carried out by an expert welder, the operation seems very simple to the inexperienced eye. Attempts to attain the same results as an expert may appear to be better than they actually are. Welders who do not realize what is taking place during the melting of the metal tend to become self-satisfied, never doubting their own ability. The presence of seemingly small defects such as undercut, porosity, lack of penetration or craters incorrectly filled may lead to failure during service if they are left in the weld. The failure of welds made in mild steel may be of a ductile or brittle nature.

The difference between a ductile fracture and a brittle fracture may be understood from the examination of two specimens, one which has broken in a ductile manner and one which has broken in a brittle manner. If both specimens are subjected to a tensile test the one which broke in a ductile manner will show a definite amount of deformation in the vicinity of the fracture. In the case of the specimen which broke in a

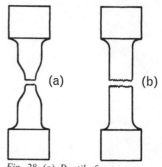

Fig. 28. (a) *Ductile fracture.*
(b) *Brittle fracture.*

50

brittle manner, there is little or no deformation in the vicinity of the fracture. (See Fig. 28(*a*) and (*b*).)

Ductile fracture

If a material is ductile it can be subjected to a tensile pull and will deform considerably before fracture takes place. The deformation of a specimen which has broken in a ductile manner may be measured after a tensile test has been carried out. This may be done by placing two centre punch-marks at a distance of 2 in apart on the specimen before testing. During the test the specimen elongates uniformly along its entire length until the ultimate tensile strength is reached. Then the specimen elongates mainly in a small local area; this is accompanied by a reduction in the diameter of the specimen within the small local area. This can be seen visually in the case of a ductile specimen. Eventually the specimen will fracture within the area of reduction. If the two broken pieces of the specimen are placed together and the distance between the centre punch-marks is measured, the distance will now exceed the original 2 in. If the new length between the centre punch-marks is found to be $2\frac{1}{2}$ in, then the specimen will have elongated $\frac{1}{2}$ in on the 2 in length. This is 25 % elongation, and is a typical example of a ductile fracture. When a ductile fracture is examined, the ends have a smooth appearance on their surface.

Brittle fracture

As seen in Fig. 28(*b*), a brittle fracture shows very little deformation in the local area around the actual break. The ends of the broken parts appear as if the parts have been sheared, with little reduction in the diameter of the specimen. Brittle fractures are extremely rapid when they occur, well within a fraction of a second.

Slipping of atoms

The question of whether a mild steel can break in a ductile manner is a question of whether the atoms of which the steel is composed are capable of slipping over each other. If the slipping of atoms is prevented from occurring, the mild steel will break in a brittle manner. Regardless of how the atoms are prevented from slipping, a brittle fracture will occur and there will be no deformation in the vicinity of the fracture.

When steel solidifies, the atoms of which it is composed form a regular pattern within each individual grain. Any single grain will have its atoms in lines or rows. These may be referred to as planes of atoms, each plane representing hundreds of thousands of individual atoms. (See Fig. 29.)

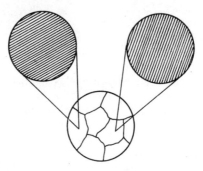

Fig. 29. Planes of atoms. Note: these may be in different directions in different grains.

If different grains are considered, the planes of atoms will be in many directions. The application of a tensile load exerts stress on all the planes of atoms, regardless of their direction. Before the tensile load is applied, the mild steel is polished and etched to

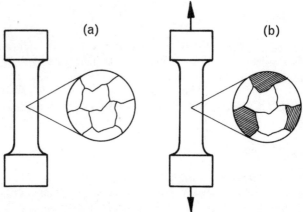

Fig. 30. (a) Steel unstressed. (b) Stressed and plastically deformed.

reveal the grain boundaries only, when viewed under the microscope. (This means that the pearlite cannot be seen.) The grains are revealed in the unstrained equiaxed condition. (See Fig. 30(*a*).)

When the tensile load is applied, the mild steel stretches and deforms elastically. While this is taking place, the atoms move a little from their original position. This means that many planes of atoms are attempting to slip over adjoining planes of atoms. Provided that the elastic limit of the steel is not exceeded, the atoms will not move over the planes of atoms below them. If the load is released, they will return to their original position. When loaded sufficiently for plastic deformation to occur, then whole planes of atoms slip over one another and cannot return to their original position. The movement or slipping of individual atoms cannot be seen with the microscope because the magnification is not high enough. However, lines indicating that slipping has occurred will appear on the polished and etched specimens. From Fig. 30(*b*) it can be observed that the angle of the slip planes are all in the same direction in any individual grain, the planes of atoms cannot overlap each other. The mild steel, having once slipped along a plane of atoms, is strengthened in that particular slip plane. Any further application of load results in slipping taking place along other planes of atoms. When the slipping of the planes of atoms occurs, the external shape of each individual grain alters in a regular fashion to the direction of loading. The grains are elongated, hence they become work hardened or, as is better known, cold worked.

It is interesting to observe the behaviour of a low carbon steel welding wire when cold and hammered continuously on an anvil. The end at which it is being hammered becomes flattened and deformation occurs. This continues as long as there are slip planes available. Once they have all been used up, then fracture results. Because the welding wire was ductile, it deformed considerably before fracture. It is worth mentioning at this stage that the hammering of steel welds when black hot is to be deplored. Some of the available slip planes will be used up when hammering and even if it did not fracture, the material will have been made more brittle.

If anything interferes with the planes of atoms slipping over one another a brittle fracture will result. Weld defects are capable of producing brittle failures. The defects may act as stress raisers, and in this capacity they set up forces which prevent the planes of atoms from slipping over each other, hence brittle fracture occurs. Consider a load applied as shown in Fig. 31(*a*). Another force will be set up at the base of the notch which has been created by lack of root penetration. The effect of this is to cause a force in another direction, thus preventing the atoms from slipping. Therefore the mild steel has stresses in two directions at once, due to the notch created, and it may now fail in a brittle manner. To understand this more clearly, the planes of atoms at the base of the notch, as shown in Fig. 31(*b*), Arrow A, slip at this point first. Slipping occurs rapidly, this being the equivalent of cold work at one concentrated point. It will be followed by a tiny crack which will spread through the mild steel plate instantly, and the failure will be a brittle one. If there had been no weld defect, the material would have deformed considerably before the fracture, which would have been a ductile one.

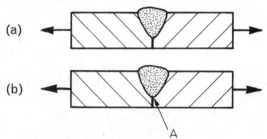

Fig. 31. (a) *Weld with defect under tensile load.*
(b) *Arrow A showing stress concentration.*

Various defects on welded joints are capable of forming stress raisers. Fig. 32(*a*) is an example of undercut acting as a source of stress concentration, whilst Fig. 32(*b*) has an external notch caused through incorrect shape of deposit. Fig. 32(*c*) and Fig. 32(*d*) are examples of lack of root penetration. In the case of the fillet weld, Fig. 32(*d*), a high concentration of stress may be put on the area marked Arrow A if any force is applied to the joint.

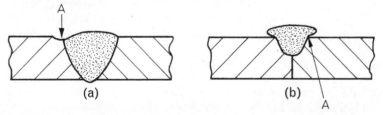

Fig. 32. (a) *Undercut. Arrow A showing stress concentration.*
(b) *Incorrect shape of deposit. Arrow A showing stress concentration.*

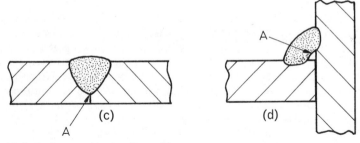

(c) *Lack of root penetration. Butt weld.*
(d) *Lack of root penetration of a fillet weld. Arrow A showing stress concentration.*

It should now be apparent that any form of defect is a potential stress raiser and even spatter, or a cut having gone just a small amount too far, could be the source of a stress concentration.

An interesting fact about certain mild steels is that they are characteristically ductile at or above say 15°C; whilst if their temperature is lowered to −6°C they may behave in a brittle manner, especially in the presence of a notch. Fig. 33 illustrates a typical brittle, unpredictable and ductile range. For this reason, low preheats are given to mild steels when at these low temperatures so that the characteristic ductility is recovered before welding.

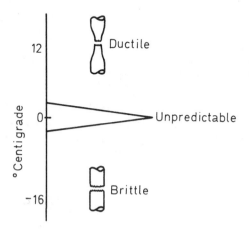

Fig. 33. Showing type of fracture expected with temperature change.

Hot cracking

This may occur when welding steels if the sulphur content is increased above 0·04 %. Impure acetylene gas, or pick-up of sulphur from any other source may be the cause of the increase. Normally steel will solidify and, as the grains are formed, any low melting point impurities are forced out to the grain boundaries. Sulphur combines with iron forming iron sulphide, melting at 980°C, and may be regarded as a low melting point impurity. If it were possible to examine microstructure of a piece of steel held at 1 000°C, the steel would be in the solid state. (See Fig. 34(*a*).) The iron sulphide at the grain boundaries is still in the liquid state at this temperature. When the temperature drops below 980°C, the iron sulphide also solidifies. (See Fig. 34(*b*).) It is now possible to consider a weld in steel solidifying. Once the steel is solid, it contracts whilst the temperature is dropping. The contraction range is considerable before the iron sulphide solidifies and the fracture occurs just before the iron sulphide becomes solid, therefore it will follow the grain boundaries. On cooling down to room temperature, the hot crack, if it extends to the surface of the steel, is easy to recognize because it will be covered in a layer of oxide.

Whilst iron sulphide has been chosen as an example of a low freezing impurity, there are many others.

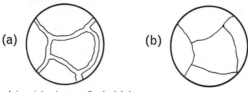

Liquid Iron Sulphide

Fig. 34. (a) *Steel held at 1 000°C.* (b) *Steel when solid at 950°C.*

A more serious source of hot cracking occurring is when welding is carried out on mild steels of heavy section where the joint is highly restrained. In these circumstances, the weld deposit may have to withstand most of the contraction stresses, therefore it must be ductile and capable of plastic deformation. If welding is carried out by the manual metal-arc process, then the electrode should be of a hydrogen-controlled type. Any hydrogen entering the weld deposit could effect the plasticity of the deposit. This, along with metal deposited which is concave in shape, and with a small throat thickness on a root run of weld, could result in hot cracking. The restraint of the parent plates tends to set up stresses in the rapidly contracting weld metal, leading to cracking. Poor joint fit-ups and incorrectly filled craters will increase the risk of cracking.

Crater cracks

When welding is carried out by arc methods, there is a tendency to leave a weld crater at the end of a run of weld. If this is not filled correctly it may tend to be concave. When cooling takes place, the crater cools rapidly while the rest of the weld is cooling slowly. The crater solidifies from all sides to the centre, therefore there is contraction from all sides to the centre, leading to possible cracking. It should be realized that once the crack has started, it may spread further into the weld seam. (See Fig. 35.)

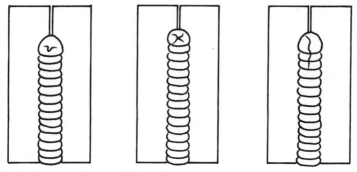

Fig. 35. Crater cracks caused by hot shrinkage.

Cold cracking

If a hardenable steel is welded without first preheating the joint, rapid cooling will cause hardening of the base material in the heat-affected zone. Circumstances favouring hardening occur when welding with the manual metal-arc welding process, which is the most popular process for the welding of these types of steels. The heat from the arc raises a small area adjacent to the weld to the austenitic condition. The rapid cooling caused by heat losses into the plates may produce hard and brittle martensite. If the stresses set up by contraction are high enough, then cracking of the base material will result. This takes place in the heat-affected zone and the crack is known as a hard-zone crack. The joint is more liable to crack if the material contains alloying elements which favour the formation of martensite. As the percentage of carbon increases, so does the chance of martensite forming, and this means that the formation of a hard zone crack is liable. (See Fig. 36.) In order to prevent hard zone cracking, a preheat of between 150°C and 350°C is applied. This preheat temperature is in the range at which austenite will transform to martensite during cooling. The effect of this preheat is to avoid the drastic quench rate and therefore allow time for the transformation to take place to softer structures, through a slower rate of cooling. Regardless of the type of electrode which is used to make the joint, it is advisable to use a flux coating which is hydrogen controlled to prevent further complications.

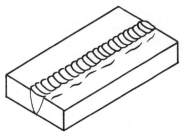

Fig. 36. Hard zone cracks.

Underbead cracking

The gas hydrogen is soluble in liquid steel and, to a limited extent, in austenite. The solubility decreases rapidly when cooling and transformation occurs to any other structure. Regardless of the rate of cooling it will be rejected out of solution. If the cooling rate is slow, as in the case of an oxy-acetylene weld, the hydrogen will have time to diffuse out to the atmosphere. When manual metal-arc welding, the rate of cooling is rapid. The result of this is that hydrogen is trapped in the material, resulting in brittleness. This means that with a hardenable steel the contraction stresses and the formation of martensite, along with the presence of hydrogen, guarantees underbead cracking. (See Fig. 37.) By the use of pre-heating, the martensite may be prevented from forming, along with less drastic contraction stresses. Therefore two of the causes of cracking are reduced. The third factor is the embrittlement caused by hydrogen. The presence of hydrogen may be reduced by the use of dry, low hydrogen-type electrodes. In addition to this, the plates to be joined should be free from moisture, oil, etc.

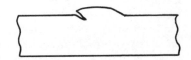

Fig. 37. Underbead cracking.

Phosphorus in steels

Phosphorus is detrimental in steels; its presence should not exceed 0·04%. If it is contained in excess of this percentage the steel will be rendered brittle, with cracking tendencies. It is generally thought that phosphorus in steels, above the permitted amount, causes cold cracking. Whilst this is true, it should be remembered that these brittle tendencies which lead to cracking exist throughout the entire range of temperature of the steel.

Porosity of weld metal

During the course of welding, defects may occur which are really minute pockets of gas. During solidification, the liquid weld metal may be releasing gases in the form of gas bubbles. If there is insufficient time for the gas to reach the surface of the weld before solidification occurs, then these gas bubbles will be trapped in the solid weld metal. (See Fig. 38.) Sometimes the porosity or blow-holes can be seen on the surface.

Porosity can be caused by hydrogen compounds formed in the molten weld pool. This gas has been described as being connected with underbead cracking. How-ever, if hydrogen has time to react

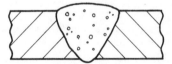

Fig. 38. Porosity.

with other elements present, such as oxygen or sulphur, compounds will form. These may be water vapour or hydrogen sulphide, and during solidification, they may be trapped as blowholes in the weld. Sulphur, in excess of 0·04% in steel, is capable of forming hydrogen sulphide, as previously mentioned, or it may react with oxygen to form sulphur dioxide. This in itself may also result in porosity. Another possible cause of porosity is the combination of carbon with oxygen forming carbon monoxide gas bubbles. The source of the carbon may be an oxy-acetylene flame which is carburizing, carbon in the weld pool, or oil or grease on the plates.

The effect of atmospheric gases on welds

By this stage it is well known that the atmospheric gases, oxygen and nitrogen, should be kept out of weld pools. During welding, if oxygen from the atmosphere is allowed to enter the molten weld pool, it reacts with the carbon first, and it may then attack the iron. On solidification, the iron oxide so formed exists at the grain boundaries of the weld metal, resulting in weakness. Iron oxide at the grain boundaries results in all the desirable properties of the materials being seriously affected. Small amounts of oxygen also have this effect. Needless to say, the oxy-acetylene cutting process is the extreme. In this case the iron is completely converted to iron oxide throughout its thickness in the kerf. The carbon reacts with the oxygen to form first, carbon monoxide, then carbon dioxide. Nitrogen introduced into weld metal from the atmosphere combines chemically with iron forming iron nitride. If the nitrogen content exceeds 0·03 %, then excess brittleness results.

From this chapter the reader will realize the importance of adequate protection of the weld from the atmosphere. This may be achieved by fluxes, slags and protective gases. Though the welder has no control over high sulphur or phosphorus content material, he is in a position to ensure cleanliness of weldable plate surfaces before welding is attempted. The composition of the material to be welded should be known, thus allowing consideration to be given when choosing a suitable welding process, correct filler material, preheat temperature, postheat temperature and welding procedure. During welding, the correct manipulation of the heat source, particularly with regard to speed of travel, is important. This allows the weld pool to solidify at a speed which will allow the gases to escape and the slag to float to the surface. It may be emphasized that the avoidance of defects such as undercut, lack of root penetration, surface lumps and craters, or anything that may be regarded as a notch, will ensure satisfactory service. If any of the above defects are present on the completed weld they will act as stress raisers which may lead to brittle failure in service.

Unfamiliar terms used in this chapter

Hardenable: Steels which may be hardened by quenching.
Deformation: To alter a material in shape.
Vicinity: The area close to the weld.
Kerf: The cut edge.
Hydrogen controlled: This refers to the flux coating around a manual metal-arc electrode. The content of hydrogen is deliberately at a minimum.

Chapter review questions

What is the difference between a ductile and a brittle fracture?
Why does cold working sometimes lead to cracking?
What are slip planes?
Is steel ductile over its entire temperature range?
Name the defects which may act as stress raisers.
Why should craters be filled up correctly at the end of a weld?
What factors may lead to cold cracking?
List the causes of hot cracking.
When welding heavy sections of mild steel, what would be the effect of highly restrained plates when depositing the first run of weld?
What are the factors leading to 'underbead cracking'?
State the causes of porosity.
Why is preheating carried out when welding certain hardenable steels?
Is hydrogen soluble in ferrite?

CHAPTER 6

Low Alloy, High Tensile Steels

Steel manufacturers have obtained very satisfactory results by adding various amounts of alloying elements to plain carbon steels. Before attempting to define what a low alloy steel may have to offer to industry, it is well to recall the composition of a plain carbon steel. It contains carbon up to 1·7%, with very small percentages of manganese and silicon added intensionally, and with impurities of sulphur and phosphorus not in excess of 0·04%. The properties of a plain carbon steel are dependent upon the *carbon* content, whether in the heat treated condition or not. Consideration should be given to Chapter 2, 'The Influence of Carbon on Iron'. If a 0·5% carbon steel is heated above the transformation range, it will change from a mixture of ferrite and pearlite to a solid solution known as austenite. A steel of this composition can be hardened by cooling it rapidly, for example, by plunging it into cold water.

A low alloy steel contains alloying elements in excess of those found in a plain carbon steel, i.e., manganese and silicon, or, as is often the case, additional alloying elements. The low alloy, high tensile steel group of alloys steels have brought about many desirable properties which may be regarded as better than those of the plain carbon steels. For example, it has been the aim of steel manufacturers to produce steel with a higher yield point and tensile strength, at the same time maintaining adequate ductility. Referring back to plain carbon steel, it will be recalled that as the carbon content increases, so does the hardness and tensile strength, but the ductility decreases. There will also be reduced impact values. Therefore the applications of the plain carbon steels are limited.

By the addition of some alloying elements, the rate of hardening will vary. Some will harden by quenching in water or oil, others by simply allowing to cool in air (not air blast). The higher alloyed steel groups, such as 14% manganese steel will remain austenitic if quenched from 1 000°C. Austenitic stainless steels containing, for example, 18% chromium and 8% nickel, do not harden at all on cooling, the structure remaining austenitic regardless of whether the rate of cooling is fast or slow. Because these types of steel do not undergo any structural changes when cooling, they do not pass through the transformation range. They cannot be hardened by heat treatment, i.e. quenching. They can only be hardened by work hardening.

The group of low alloy steels, with which this chapter is chiefly concerned, is *hardenable* by heat treatment, i.e., quenching. When certain alloying elements are present such as chromium, manganese, molybdenum, tungsten, along with the effect of the carbon, the normal change from austenite is delayed during cooling. The change from austenite does not take place at about 700°C but occurs at a lower temperature; this is quite acceptable so long as there is sufficient time for the change to take place. Should the cooling rate be rapid then the austenite does not have time to change to ferrite and pearlite, but it changes directly to hard martensite.

When welding is carried out without preheat on low alloy steel plates, hardening occurs in the heat-affected zone, because of the rapid cooling rate. It must still be kept in mind that, despite the hardening of the base materials, the original intention of the manufacturers was *not* to increase the hardness. It was to increase the ultimate tensile strength, the yield point and, at the same time, maintain reasonable ductility. An example of the correct mechanical properties being required is when a low alloy, high tensile steel is used for structural purposes. If these give way to hard brittle structures which are liable to cracking or may be already cracked, then the results will be disastrous.

Composition of low alloy steels

The total of the alloying elements used does not usually exceed 5%. These steels should not be confused with the higher alloy steel groups such as 18/8 austenitic stainless steels. There are very many different alloy steels and it would be impossible to list them in anything but general terms. The grades particularly suited to

Welding Science and Metallurgy

welding contain small percentages of carbon, usually below 0·15%. Carbon has far more influence upon hardening of the base material than any other alloying element. Despite the low carbon content, hardening of alloy steels may still take place through the influence of other alloying elements. The low alloy, high tensile steels may contain chromium, manganese, molybdenum, nickel and tungsten. A typical example of an alloy steel which has good welding properties is 0·1% carbon, 0·7% manganese, 0·3% nickel and 0·3% molybdenum. A popular alloying element is molybdenum, but it is not usually used alone in low alloy steels. The presence of molybdenum in steel has the effect of reducing the brittleness throughout the entire range of cooling, therefore assisting in the retention of ductility. Because of this, the cracking risk associated with the welding of low alloy steels is reduced. The composition of low alloy, high tensile steels varies considerably, but whatever the composition, molybdenum is usually included in the filler material, i.e., electrode. It is fortunate that the alloying elements in a low alloy steel are not 'poison' to each other. Therefore the low alloy steels may be welded satisfactorily using an electrode which contains different alloying elements to those of the parent material, the weld deposit having properties that compare favourably.

When welding low alloy steels, consideration should be given to the following:

Choice of welding process

Oxy-acetylene
This is usually considered a very slow process, leading to excessive overheating of the low alloy steel being welded. Its applications are usually confined to light gauge sheets and small diameter sections. When used, a low alloy filler rod is desirable.

Tungsten inert gas
High quality joints are made economically on thin sheets of low alloy steels. This process allows excellent concentration and control of heat, therefore a high quality weld is produced with a limited total heat input, keeping distortion to a minimum.

Metal inert gas and carbon dioxide shielded arc
On thicker sections the metal inert gas process may be considered. The inert gas shield around the arc prevents contamination from

the atmosphere and discourages losses of alloying elements which may react with oxygen forming gaseous compounds. A suitable filler material, not necessarily of the same composition as the material being welded, is used. It is important that the filler wire is in a deoxidized condition.

When carbon dioxide is used for welding low alloy steels, the composition of the shielding gas allows some oxidation of the alloying elements present in the material being welded. This can be balanced to some extent by using a deoxidized filler wire. In general, this process is less flexible, and changing over from welding one low alloy steel to welding another may radically change the qualities attained.

Manual metal arc

Excellent results are obtained by the use of this process. This process is the most widely used of all for the welding of low alloy, high tensile steels. Because this process is the most widely used, it may be considered in detail. Most of the problems which are associated with the manual metal-arc process apply equally to other processes.

Preheating

Some low alloy, high tensile steels harden when they are heated to a high temperature and cooled rapidly. When these steels are welded, hardening and embrittlement occur in the heat-affected zones on either side of the weld. Unless preheating is carried out, the localized hardening can be serious in the case of low alloy, high tensile steels, because of the possibility of cracking during cooling. These cracks may be the result of shrinkage stresses during cooling, the shrinkage stresses being strong enough to result in cracking of the hard, brittle constituents in the heat-affected zone. A slower rate of cooling allows the stresses to be more evenly distributed throughout the heat-affected area.

Many weld failures have resulted from welding low alloy steels without preheating. It is dangerous to think that the first layer of weld deposited does the preheating for subsequent layers, and that any cracks occurring in the first layer or bead may be burned out during the deposition of the second layer. It is impossible for the arc to penetrate deep enough to melt out any cracks which are present in the root bead.

65

When welding low alloy, high tensile steels the preheat temperature is usually between 150° and 300°C. The preheat temperature varies with the thickness of the material being welded, and also with the composition. A guide to the correct preheat temperature is that when the thickness exceeds $1\frac{1}{2}$ in, then the higher preheat temperature within the range is applied. It is not sufficient to heat the steel to the proper preheat temperature; the temperature must be maintained until the weld is complete.

Test for correct preheat temperature

Usually, the correct preheat temperature will be supplied by the manufacturer of the particular low alloy steel to be welded. However, the chemical composition of the low alloy steel may not always be known or it may be necessary to establish whether or not a preheat is necessary. A quick testing method used in workshops is the clip test. (See Fig. 39.)

This test is applied to alloy steels of $\frac{3}{8}$ in or over in thickness. The clips or lugs to be welded on to the steel under test are low carbon steel plates, 2 in square and $\frac{1}{2}$ in thick. The clip is fillet welded to the unknown steel, as shown in Fig. 40(*a*). The weld is made with the size and type of electrode and the welding current and speed that will be used for the welding of the unknown steel component. Allow the weld to cool until room temperature is

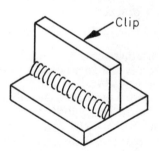

Fig. 39. Clip test for determining preheat temperature.

reached and then break the clip off by hammering. If the lug bends and, after a number of blows, falls through the weld, the test indicates that the composition of the steel, the preheat temperature of the steel when used and the welding procedure adopted are such that underbead cracking is absent. (See Fig. 40(*b*).)

A steel that requires an increase in the preheat temperature will fail the clip test. This is indicated by some of the hard brittle parent metal being pulled out, as shown in Fig. 40(*c*). The fracture occurs in the base material because of the hardened condition of the heat-affected zone. It may be necessary to repeat the test several times

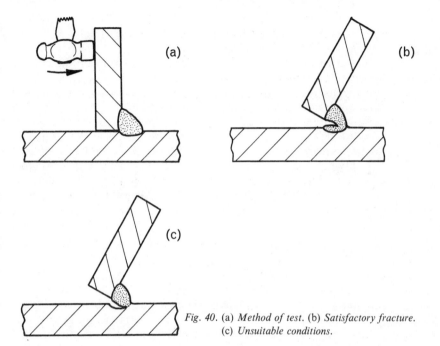

Fig. 40. (a) *Method of test.* (b) *Satisfactory fracture.*
(c) *Unsuitable conditions.*

using a higher preheat temperature, up to a maximum of 300°C. If underbead cracking still occurs, then it is necessary to reconsider the choice of electrode and the welding procedure being followed.

Choice of electrode

The flux coatings on electrodes which are used for welding of low alloy, high tensile steels are of a hydrogen-controlled type. This is to prevent hydrogen entering the weld metal and causing embrittlement. During the cooling of a steel weld, the solubility of hydrogen in the weld metal decreases rapidly. With slow cooling, most of it diffuses out to the atmosphere, if cooling is rapid then it has insufficient time to diffuse to the atmosphere and is trapped in the crystalline structure of the solid steel. It should be noted that some of the hydrogen may be forced into the heat-affected zone. If this does occur, then the effect of rapid cooling, producing martensite, along with contraction stresses and the presence of hydrogen, will result in underbead cracking. No doubt the reader will have guessed correctly that a preheat would have avoided the conditions leading to underbead cracking.

It is important that a low hydrogen type of electrode should be kept dry. There is no such a thing as a damp, low hydrogen electrode; it loses that title when damp, because, at arc temperatures, the moisture splits into hydrogen and oxygen.

Alloys used in low alloy, high tensile electrodes must have superior welding properties. Molybdenum, for instance, is one of the best alloying elements for the core wire of welding electrodes. Very little of the molybdenum combines with the slag or is lost by oxidation. The molybdenum steels have a good combination of ductility and strength throughout the temperature range, from solidification down to room temperature.

Tacking

Tacking should not be carried out on cold plates. This would produce a hard brittle zone which may crack even before welding commences. Therefore tacks should only be made on preheated work. The fact that low alloy, high tensile steels, as the name suggests, are of high tensile strength and are capable of setting up very high contraction stresses means that distortion will occur. To control distortion, tacks should not be less than about 4 in in length, and they should be made with a large diameter electrode with a high current value. On these types of steels, the tacks should always be of two layers.

Welding procedure

Reference has been made to the importance of preheating, choice of a suitable electrode, with its correct flux coating and core wire composition. With the parts suitably tacked in position, welding may now proceed. The deposition of heavy beads are less susceptible to cracking than the deposition of small beads. At the same time, distortion will be less. To maintain the heat in the joint during welding, a 6 s.w.g. flux-coated electrode should be regarded as a minimum for most low alloy, high tensile steel fabrications. To avoid distortion, which may be worse than when welding a similar size mild steel fabrication, it is necessary to apply the use of recognized weld sequences and other means. Backstepping, skip welding, presetting and also the use of jigs, fixtures and manipulators are desirable.

Unfamiliar terms used in this chapter

Hardenable: To increase hardness by heating and quenching.

Heat-affected zone: The area on either side of a weld where the structure is affected by the heat of welding.

Deoxidized: In this case, the materials concerned have had all traces of oxygen removed.

Hydrogen controlled: A flux coating which contains the minimum amount possible of hydrogen.

Chapter review questions

Which elements may be used in low alloy steels?

What are the effects of alloying elements upon the mechanical properties of the steel?

Give the condition of the heat-affected zone when a low alloy steel has been welded with (*a*) no preheat, (*b*) a suitable preheat.

What are the relative merits of the various welding processes which are available for the welding of low alloy, high tensile steels?

Which alloying element is desirable in a flux-coated, manual metal-arc electrode used for low alloy, high tensile steels?

Why is it important that there should be no cracks left in the first layer?

Give the usual range of preheat temperatures for alloy steels.

What is the purpose of a clip test?

Name the probable structure to be found if there is insufficient time on cooling for ferrite and pearlite to form.

Why is it important that damp electrodes should never be used?

Explain the effect of hydrogen when it is trapped in a welded fabrication.

What precautions should be taken when welding low alloy, high tensile steels?

CHAPTER 7

Stainless and Heat Resisting Steels

The ease with which iron and carbon steels oxidize is widely known. Cutlery manufactured from these materials was susceptible to rusting and was therefore unhygenic. To prevent this rusting, a stainless type of steel was introduced. This alloy was composed of iron, carbon and chromium and was able to resist oxidation.

Nowadays, in addition to iron, carbon and chromium, certain alloys contain varying amounts of nickel.

Effects of alloying

Chromium

A feature of adding chromium is the formation of a very thin, invisible layer of oxide which tightly adheres to the surface, therefore protecting it against further oxidation. The thickness of this oxide film increases as the temperature of the alloy is raised. Temper colours formed on the surface during heating reveal the presence of this film. In the presence of air, stainless steel, when near or above its melting point, forms the refractory chromium oxide which is visible as a black scale. Chromium added to iron forms a solid solution making the ferrite stronger at all temperatures and increasing its corrosion and scaling resistance. Mechanical properties such as hardness, strength and resistance to wear are increased in steel by the addition of chromium.

Chromium has a strong affinity for carbon, with which it may combine to form chromium carbide, which is extremely hard. The hardness of this compound can be appreciated when it is realized that chromium carbides are used for hardsurfacing weld deposits.

Nickel

Unlike chromium, nickel does not form carbides but forms a solid solution with the iron. Very favourable properties are obtained from the addition of nickel; these include increases in hardness, strength and ductility, so that there is an increase in the toughness of the alloy. When the nickel content in steel exceeds 6%, there is no transformation on heating or cooling, the structure being austenite at all temperatures below the freezing point.

Thermal conductivity

The thermal conductivity of stainless steel is much lower than that of a mild steel. Because of this, stainless steels are more susceptible to local overheating and distortion problems during welding.

Electrical conductivity

This is quite low, which is advantageous when resistance welding processes are being used.

Expansion and contraction

The expansion rate of stainless steel is approximately one and a half times greater than that of mild steel. When welding, this results in excess shrinkage stresses, thereby causing distortion.

It should be appreciated that the aforementioned properties will vary with the composition of the alloy. The higher the chromium content, then the more pronounced are the properties. The high chromium steels are classified as martensitic, austenitic and ferritic. These types may be considered individually as follows:

Martensitic stainless steel

The composition of this group is from 11·5% to 14% chromium and from 0·2% to 0·4% carbon, the base metal being iron. This type of stainless steel has air hardening tendencies; martensite is formed regardless of the rate of cooling from high temperatures. When welding is carried out on this type of stainless steel, the difficulties which arise are high hardness, accompanied by brittleness, which may result in cracking. The problem may be overcome by correctly preheating; this should be to about 350°C. Advantages to be gained from preheating are a reduction of hardness, thereby reducing cracking tendencies, and a considerable reduction in the

71

shrinkage stresses because of the temperature gradient being reduced. The deposited weld metal should preferably be of a ductile character; the use of a covered electrode containing 25 % chromium and 20 % nickel meets this requirement. It is important to realize that, on completion of welding, the heat-affected zone will be the weakest area when a nickel chromium alloy electrode has been deposited, therefore postheating is recommended immediately after welding. A postheat temperature of about 750°C will ensure an acceptable amount of ductility.

Ferritic stainless steel

The composition of ferritic stainless steels is from 16 % to 30 % chromium with a maximum of 0·1 % carbon, the base metal being iron. This material is also referred to as stainless iron or chromium iron. It has a high resistance to corrosion and is not heat treatable. Very little austenite is formed when it is heated to a high temperature and it may be regarded as non hardenable. This has been stated regardless of the 0·1 % carbon content and the reader should appreciate that this small amount is insufficient to produce any notable increase in hardness. However, this small amount of carbon would change to austenite during welding. The heat-affected zone on each side of the weld would undergo considerable grain growth and the small amount of austenite formed might find itself entrapped within the grains. Unless the rate of cooling is correctly controlled, the intergranular austenite, on cooling, will revert to martensite, with little change in the hardness but there will be embrittlement of the alloy. (See Fig. 41(*b*).)

(a)

(b)

Fig. 41. (a) *Ferritic stainless steel.*
    (b) *After welding. Note: there is martensite in the grains and at the boundaries.*

Welding may be successfully carried out on this type of alloy if a low preheat of 150°C is applied. With the manual metal-arc welding process, use is made of an electrode containing 25% chromium and 20% nickel, which ensures a ductile weld deposit. Sometimes a combination of electrodes is used, the weld layers being made with the 25% chromium, 20% nickel alloy electrodes. These are deposited as shown in Fig. 42, the final layers being of a high chromium content, the electrode containing from 26% to 30% chromium. This will give a surface that is highly resistant to corrosion, i.e., acid. The electrodes, in addition to the normal alloying elements, should contain a small amount of nitrogen to help in preventing an increase in the grain size. A postheat should be carried out immediately after welding, to a temperature of 730°C, to control the embrittlement tendencies.

Fig. 42. Showing order of deposition using two different types of electrodes.

Austenitic stainless steel

These are the most popular types of stainless steel used industrially. Fortunately they are readily welded by many processes.

Chemical composition

The composition of austenitic stainless steels varies considerably. Chromium is present from between 7% and 30%, whilst the nickel content ranges from 6% to 36%. The carbon content rarely exceeds 0·25%, the base metal being iron. For industrial purposes, alloys containing around 18% chromium and 8% nickel are widely used.

Mechanical properties

The tensile strength of an 18/8, i.e., chromium-nickel steel, is between 40 and 50 tons per square inch, the hardness is between 160 to 200 Brinell. Its ductility is also high, the elongation percentage being from 40% to 60%.

F

Physical properties
The melting range varies, depending upon the chemical composition. An 18/8 alloy melts between 1 420°C and 1 450°C. In general, the thermal and electrical conductivity is low, whilst the expansion rate is quite high. Perhaps the outstanding property of austenitic stainless steel is its ability to resist oxidation, corrosion and heat.

Chemical properties
The 18/8 austenitic stainless steel takes the form of a solid solution if quenched from 1 000°C or slow cooled. Because there is no transformation on heating and cooling, it is not subject to hardening by quenching.

While it has been stated that an austenitic stainless steel is oxidation resistant at room temperature, layers of chromium oxide will form at elevated temperatures. It should be stated at once that any oxide layers will ruin any attempt at welding, regardless of the process, so that the oxides must be removed and prevented from re-forming. For example, when manual metal-arc welding, it would be impossible to maintain an arc with a bare wire electrode because of refractory oxides forming. Therefore a flux-covered electrode is essential. Fluorides are contained in the flux coating for the purpose of removing any oxides which may form during welding.

Cutting of austenitic stainless steel cannot be carried out by the oxy-acetylene process. The reason for this is that the refractory chromium oxide formed during heating prevents the cutting action from taking place. This difficulty has been overcome by using powder injection cutting processes.

Austenitic stainless steels have one undesirable property. When a weld is made between two pieces of this material, heat is lost by conduction through the plates from the weld area. (See Fig. 43(*a*).) The temperature gradient along the plates on either side of the weld contains an area which will remain at a temperature of between 750° and 450°C for a considerable period of time. (See Fig. 43(*b*).) The areas which are held within this temperature range undergo an internal structural change which is most detrimental to the corrosion resisting properties of the material. Whilst held at this temperature, chromium combines with carbon, forming chromium carbide. Once formed, this is precipitated out of the austenitic solid solution into the grain boundaries. This is indicated on the microstructure in Fig. 43(*b*).

The chromium is taken from the layer of metal immediately adjacent to the grain boundary, thus leaving the ferrite in this area

Fig. 43. (a) *Heat losses from the weld area.*

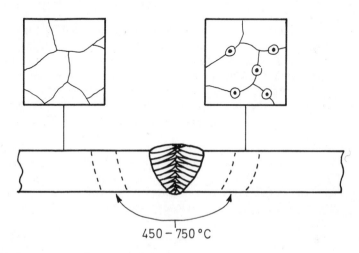

450 – 750 °C

(b) *Areas prone to chromium carbide precipitation.*

devoid of chromium. In turn, this reduces the resistance of these areas to corrosion. If these areas are subjected to corrosive conditions in service, intergranular corrosion now takes place. Once intergranular corrosion has started, it continues until the ultimate failure of the material occurs. (See Fig. 44.) Another name for intergranular corrosion is weld decay. This is a misnomer, the failure occurring alongside the weld but not actually in it.

Fig. 44. Failure through intergranular corrosion.

Intergranular corrosion may be prevented by at least three different methods:

(1) Weld decay may be prevented by reducing the carbon content to such a small amount that carbide of chromium cannot form and therefore no precipitation can take place. The carbon content must not be in excess of 0·03 %. These steels are very suitable for corrosion and oxidation resistance applications, but should only be used for applications where the temperature does not exceed 450°C.

(2) A second method is simply to heat the material to 1 000°C after welding, followed by quenching in water. The rapid rate of cooling ensures that cooling occurs so quickly through the precipitation temperature range that there is insufficient time for any precipitation of chromium carbide to take place. This method is theoretically easy, but unfortunately the parts which have been joined may be much too large for this to be a practical proposition.

(3) Another method is the addition of alloying elements which have a greater affinity for carbon than chromium. This means that the chromium remains unaffected, still remaining in solution with the ferrite. The usual elements which are used for this purpose are titanium and niobium. These are known as stabilizing elements and they are carefully added by both steel and electrode manufacturers. It is usual to add the stabilizing element in proportion to the carbon content. For example, an austenitic stainless steel plate may be stabilized by adding titanium at four times the carbon content of the alloy. When manual metal-arc welding, the covered electrodes are not stabilized with titanium. The reason for this is that titanium transfers badly across the arc and it will not reach the weld pool. Niobium is used as a stabilizer in electrodes, and transfers across the arc with negligible losses. The amount of niobium required is ten times the carbon content.

The welding of austenitic stainless steel

Austenitic stainless steels are readily weldable by most welding processes. The preferred welding processes are metal inert gas, tungsten inert gas, and manual metal-arc. Regardless of which process is used, no grain refinement takes place as subsequent runs of weld are deposited. It will be recalled that when welding mild steel, the first run of weld deposited has a coarse structure. This is corrected by it being reheated when a subsequent run of weld is deposited. During reheating, the mild steel is raised in temperature through the transformation range and recrystallization of the lower deposit takes place. This action is repeated with each subsequent deposit. However, with austenitic stainless steel, there is no transformation range on heating or cooling. The first run of weld has a coarse structure. (See Fig. 45(*a*).) Any subsequent runs of weld reheat the lower runs of weld, but there is no change in the structure which remains coarse throughout. (See Fig. 45(*b*).) Hot or cold working, i.e., peening between each run of weld results in grain refinement.

Fig. 45. (a) *Coarse run of weld.* (b) *No refinement with subsequent runs of weld.*

If the austenitic stainless steel has been sufficiently cold worked, then some refinement of the grain structure will take place in the heat-affected zone during welding.

It may be summarized that successful welding by any process requires serious consideration with regards to oxidation, carbon pick-up during welding and distortion.

Oxidation

This is prevented by a flux shield in manual metal-arc welding, or by an inert gas when metal inert or tungsten inert gas welding. Quite often a backing gas of argon is passed along the underside of the joint during welding, resulting in a smooth undersurface on completion.

Carbon pick-up

This increases the risk of forming undesirable carbides during welding. Care should be taken to ensure that there is no oil or grease or anything that will give off carbon at welding temperature. Carbon dioxide should not be used as a shielding gas when welding with either of the gas shielded welding processes.

Distortion

When welding austenitic stainless steel, tacking or jigging must be carried out correctly. The high expansion and contraction rate may cause serious warping of the components. Jigging is an ideal way of holding and locating the joint correctly, and it also offers the advantage of supplying a means of applying a backing gas. If tacking is carried out, it should be closer than for the same thickness of mild steel and longer in order to prevent it from cracking.

After treatment

All flux residues and temper colours, which are oxides, should be removed after welding if maximum protection against corrosion is required. A 10% nitric acid solution is usually sufficient to remove the oxides, although the assistance of a suitable abrasive is found to be effective.

Weld processes

Austenitic stainless steel is one of the materials on which a single joint may be welded to advantage by two different welding processes. The root run of a weld may be deposited on to a pipe by the tungsten inert gas welding process; this gives a controlled amount of penetration. Further deposits may then be made with the manual metal-arc process, building up the remainder of the pipe to its correct thickness. The use of the two processes is an economical proposition.

The tungsten inert gas welding process is ideal for thin sections, or where a controlled amount of penetration is required. The metal inert gas process finds advantages on thicker sections, such as above $\frac{1}{8}$ in in thickness. Welds are easily carried out in all positions by this process. The use of small additions of other gases added to the argon assist in giving very acceptable welds. The mode of metal transfer across the arc is in the form of a spray. The smooth transfer of metal across an arc in an atmosphere of argon ensures that there is very little loss of the alloying elements. The advantage to be gained by this is that either titanium or niobium may be used for stabilizing the weld deposit. In an argon atmosphere, both titanium and niobium can transfer across an arc.

The manual metal-arc welding process is still very popular, being widely used when welding stainless steels. For consistently successful results, several points should be watched, in addition to those common to other processes. The flux-covered electrodes should be kept dry, otherwise porosity in the weld deposit will result. Austenitic stainless steel is a poor conductor and, because of the risk of overheating the electrode during welding, the manufacturers make the electrodes shorter than mild steel electrodes. Electrodes may also be overheated if the manufacturers' recommended current is exceeded. During the course of welding, a short arc should be maintained in order to prevent losses of alloying elements.

Heat resisting steels

For many industrial applications, materials are required which are capable of withstanding high temperatures. Whatever material is chosen, it must be capable of the following:
  (*a*) It must retain sufficient strength at its elevated temperature.
  (*b*) It must be creep resisting. This means that it must not deform plastically when held for a long period of time at stresses lower than the yield strength.
  (*c*) It must not suffer from oxidation at its working temperature.
  (*d*) It should be capable of being welded.
The austenitic stainless steel group fit very well into the heat-resisting classification. Use is also made of the low alloy steel group if the alloying elements chromium and molybdenum are included in small quantities.

Austenitic stainless steel

Modern heat-resisting steels may contain anything from the familiar 18/8 chromium and nickel composition to quantities of chromium between 12% and 25%, and nickel content between 12% and 28%. An essential difference between this group and the conventional stainless steels is the addition of other alloying elements. The purpose of these is to ensure that the strength of the alloy is acceptable at elevated temperatures. The most commonly used alloying elements are molybdenum, tungsten and titanium. Even the carbon content may be increased up to 0·4%. The increased amount of carbon produces carbides, resulting in an increase in the creep resistance of the alloy. The effect of tungsten carbide is to give strength at high temperature, whilst molybdenum does great service in ensuring that the alloy is creep resistant, quite apart from imparting other desirable properties. By far the most important alloying element in heat-resisting steels is chromium. The presence of chromium prevents any undesirable scaling of oxide and also assists in retaining appreciable strength of the alloy at high temperatures. Certain alloys in this group are capable of having high strength values at a temperature of over 800°C.

Welding of these alloys may be carried out in a similar manner to the welding of austenitic stainless steels. The types which have additional alloying elements for high temperature service are more prone to hot cracking than conventional austenitic stainless varieties. When welding is carried out, the manual metal-arc process may be used to advantage. This is because narrow weld beads may be deposited in a fast manner, thereby ensuring quick cooling through the range of temperature where cracking may occur.

Low alloy steels

There are thousands of varieties of low alloy steels. Some of these are alloyed with chromium and molybdenum. The chromium will rarely exceed 8%, whilst the molybdenum may be between 0·5% and 1%. These types of steel are useful for a wide range of applications where the maximum working temperature does not exceed 450°C. Welding may be carried out as with most low alloy steels. (See Chapter 6.) The main difficulty is that hardness occurs in the heat-affected zone regardless of whether a filler material similar to the parent metal, or an austenitic type of filler material is used. Correct heat treatment will assist in producing satisfactory welded joints.

Unfamiliar terms used in this chapter

Refractory: Does not melt easily.
Solid solution: Two or more elements being dissolved in one another in the solid state.
Intergranular corrosion: Oxidation around the grain boundaries.
Fluoride: Part of the composition of a flux, this removes oxides.
Affinity: Mutual attraction.
Stabilizing elements: To make the solution of iron and chromium remain.
Carbides: Compounds of a metal and carbon.

Chapter review questions

What is the purpose of chromium in stainless steels?
Compare the properties of stainless steel with those of mild steel.
Name three distinct types of stainless steel.
What is the composition of stainless iron?
By what means may stainless steel be cut?
What is intergranular corrosion?
How may intercrystalline corrosion be prevented?
When manual metal-arc welding austenitic stainless steel, why are the electrodes stabilized with niobium?
What condition will the second run of weld be in when three runs of weld are deposited—columnar or equiaxed?
Give another term for temper colours.
State the advantage of an argon backing when welding stainless steel.
When are two different welding processes used to advantage when welding stainless steels?
How is weld metal passed across the arc when metal inert gas welding stainless steel?
Why are electrodes for the welding of stainless steel made shorter in length when using the manual metal-arc process?
Give four desirable properties of heat-resisting steel.
Why are alloying elements added to heat-resisting steels?

CHAPTER 8

The Factors affecting the Welding of Cast-iron

The name of cast-iron is given to alloys of carbon and iron in which the carbon content is between 2% and $4\frac{1}{2}$%. In the majority of cases, the cast-iron used is between 3% and 4%. The carbon may be present in very different forms: either combined with the iron, in which case it forms cementite, or dissolved in the iron in the form of martensite. In each case it gives the alloy great hardness. It may also exist in the free state in the form of particles of graphite; these in turn may be in the form of flakes or spheroids.

All forms of cast-iron contain, in addition to iron and carbon, the elements, silicon, manganese, sulphur and phosphorus. For special reasons, other elements are often added to cast-irons, such as copper, nickel, chromium; when these are added the cast-iron is known as *alloy cast-iron*.

The various forms of cast-iron may be taken to be white cast-iron, grey cast-iron, spheroidal graphite cast-iron and malleable cast-iron; each will be considered separately.

White cast-iron

The carbon is in the combined state as cementite or martensite, giving an extremely brittle and hard material. The name white cast-iron is derived from the appearance of the fracture of the material which is white in colour. It may be produced by rapid cooling so that the carbon has not time to separate out. Another way in which it may be produced is by adjusting the composition of alloying elements so that carbon is prevented from forming graphite. In welding operations, it should be the aim of the welder to avoid producing white cast-iron unless specially intended for hardsurfacing purposes. White cast-iron should be regarded as a poor welding material. It is not usually arc welded, and successful welds are limited when the oxy-acetylene process is used.

Fig. 46. White cast-iron × *100.*

' **Grey cast-iron**

This is the most commonly used of the cast-irons. It has a chemical composition of between 2% and $4\frac{1}{2}\%$ carbon, 3% silicon and small amounts of manganese, sulphur and phosphorus, the base metal being iron. Grey cast-iron is so called because of the grey appearance of a fracture of a piece of this material.

Grey cast-iron has a high strength in compression—around 50 tons per square inch; its tensile strength, however, is quite low—about 14 to 18 tons per square inch. If grey cast-iron is subjected to impact testing, the values revealed are quite low.

A comparison with white cast-iron will show that grey cast-iron is much softer and readily machinable. The reason for this is that all the carbon is not in the combined state, as is the case with white cast-iron, but may be found in the form of free graphite as well as in the combined form as cementite. If a microstructure of a piece of grey cast-iron is examined, it is found that the free graphite is in the form of flakes in the main structure, which is composed of pearlite. (See Fig. 50.)

It is now necessary to consider the circumstances which may lead to the precipitation of graphite into the cast-iron. The addition of silicon to cast-iron results in the formation of ferro-silicon, which has the effect of assisting the precipitation of the carbon in the form of graphite. It is the compound cementite which breaks down into free graphite and iron. Regardless of any other circumstances, slow cooling is essential for the breakdown to graphite to occur. The temperature at which the breakdown takes place is between 750° and 650°C when cooling. Between this temperature range a definite expansion of the casting will take place. It should be realized that rapid cooling of cast-iron, whether it contains silicon or not, keeps the carbon in combination with the iron, which, in turn, results in the formation of white cast-iron. The element manganese is added to cast-iron to combine with the sulphur, forming manganese sulphide. The sulphur is an ever present impurity and should never exceed 0·15%. Both manganese and sulphur, as individual elements, tend to prevent the breakdown from cementite to graphite. When the compound of manganese sulphide is formed, the effects of the individual elements are nullified and there is no interference with the graphite formation. The small amount of phosphorus present in cast-iron, which rarely exceeds 0·75%, has no detrimental effects on the carbon in grey cast-iron. It does, however, improve the fluidity of the cast-iron.

With the knowledge of the above effects, it is now possible to consider general conditions which lead to a weld being carried out, resulting in the formation of a correctly formed grey cast-iron, both in the weld metal and the remainder of the casting. This means that the completed weld and the rest of the casting may be filed, machined or shaped. It should also possess the desirable properties associated with grey cast-iron.

Problems when welding grey cast-iron

The difficulties which may be encountered during the welding of grey cast-iron are:
(*a*) Cracking.
(*b*) Oxidation.
(*c*) Loss of alloying elements during welding.

Cracking
It must be realized that cast-iron is inherently brittle. Therefore it is particularly sensitive to the effects of expansion and contraction, especially when the heat source is localized, as in the case of oxy-acetylene or electric-arc welding. By expanding the whole or part of the casting before welding, the expansion and the contraction stresses may be evenly balanced, therefore preventing cracking. This may be done by preheating to a suitable temperature, depending upon the welding process used. If the casting is complicated in shape, the entire casting may be preheated to about 700°C, the preferred process of welding being oxy-acetylene. After the completion of welding, postheating is essential, for it will remove internal stresses which may be present as a result of welding, and will also ensure a slow even rate of cooling, providing a correctly formed grey casting throughout.

84

Oxidation

Cast-iron rapidly oxidizes at elevated temperatures. The formation of this oxide interferes with the progress of the weld. As it prevents the combination of the molten metal and also burns out the carbon, thus being conducive to the formation of white iron, it must be destroyed. The heat of the oxy-acetylene flame results in the volatilization of the silicon. To obtain good welds when using the oxy-acetylene process, they should be completed in the fastest possible time. At the same time a suitable flux should be used to dissolve the oxides formed during welding.

Loss of alloying elements during welding

The loss of alloying elements during welding may be detrimental to the quality of the weld metal. Therefore filler rods containing an excess of silicon are used. The purpose of this is to ensure that there is always a sufficient amount of silicon present in the weld, despite the losses of this element during welding. Some of the silicon acts as a deoxidizer, thus preventing blowholes, but in this role it is lost from the casting.

When welding grey cast-iron, the best results are obtained by using the oxy-acetylene process. This is the only welding process which is capable of providing a grey cast-iron structure in the weld metal, the heat-affected zone and the casting. To obtain grey cast-iron on completion of the weld, it is necessary to consider the following: preheating, neutral flame, filler rod and flux of suitable composition, correct welding technique and subsequent postheating.

Manual metal-arc welding of grey cast-iron

Many successful welding operations are carried out on grey cast-iron by this process. Quite often the preheat temperature is very low or even dispensed with altogether. Remembering that grey cast-iron is brittle, the following points should be observed before welding:

(*a*) Low heat input.
(*b*) Choice of electrode.
(*c*) Electrical conditions.
(*d*) Buttering layer.
(*e*) Skip welding and peening.

Low heat input

Strictly speaking, it is possible to preheat fully the casting before manual metal-arc welding; if this is possible then the best method would be to use the oxy-acetylene process. However, a weld may be carried out by the manual metal-arc process but the weld metal deposited would not be of cast-iron; the casting would be of grey iron throughout because of slow cooling. For most purposes, when cast-iron is welded by the manual metal-arc process, the heat input will be localized and the preheat dispensed with. Quite often preheating is not a practical proposition, for example, on a broken part of a large casting which is difficult to dismantle.

When welding, perhaps the most important point to remember is that if weld metal is deposited on to a large piece of cold cast-iron, the casting close to the weld will be cooled very quickly. The base metal just below the fusion zone will have been heated enough to result in some solution of the graphite taking place at high temperatures. With rapid cooling, which may be caused by heat losses through conduction into the casting, martensite may be formed because of the quench effect. By welding without preheating, a narrow, hard, brittle martensitic zone is produced. This may be compared with glass which is considered a brittle, hard material, but, when thin, may be quite flexible. Providing that the martensitic zone is kept narrow it is acceptable. During the course of welding, the deposition of additional beads of weld has the effect of tempering the martensite, resulting in a reduction in hardness and brittleness.

Choice of electrode

The electrodes used when welding cast-iron may be nickel steel alloy, phosphor bronze or soft iron. It will be noticed that there is not a cast-iron type of electrode composed of cast-iron. This is because serious difficulties would be encountered when attempting to weld, owing to alloying elements being lost when in transit across the arc. Because these elements are burned out, the weld would be porous, brittle and incapable of withstanding its own contraction stresses without fracture. When consideration is given to the choice of electrode most suitable for welding cast-iron, it should be realized that most metals will combine with carbon forming compounds. The result of this is the formation of a hard carbide. Nickel does not combine with carbon to form a carbide, therefore the nickel alloy electrode is suitable for welding cast-iron, producing a deposit which is soft and ductile. A phosphor bronze electrode will also be acceptable, providing there is no objection to the colour difference, and as long as the casting is not subjected to temperatures above 300°C. When a soft iron electrode is considered and is deposited directly on to the surface of a casting, dilution with the base material takes place and the carbon content of the deposit is increased. If this increase is only 0·4% (it may be much higher) any quench effect caused by rapid colling results in hardness and cracking takes place. To obtain satisfactory results using soft iron electrodes alone, a full preheat and postheat should be carried out.

Another consideration, when choosing electrodes, is their diameter. Best results are obtained by using small gauge electrodes, for example, 10 s.w.g. The reason for this is that there will be less heat input, less contraction stress and therefore less disturbance of the base material than there would be if larger diameter electrodes were used.

Electrical conditions

The choice of welding equipment is important. A minimum of heat is required when welding cast-iron without preheat. This may be controlled by use of a direct current, making the electrode positive. The advantage to be gained from making the electrode positive is that two-thirds of the heat will be at the positive pole, and therefore in the electrode, and a minimum amount of heat will go into the work which is the negative pole. By observing these conditions there will be a minimum amount of penetration ensuring that there is a minimum amount of disturbance of the base material.

The current required for a nickel alloy electrode is lower than for a similar gauge of a soft iron electrode. An example of this is the use of a 10 s.w.g. nickel alloy electrode, when correct penetration can be obtained with a current value of 80 amp; this also provides a desirable soft arc. In contrast to this, a 10 s.w.g. soft iron electrode requires 110 amp, thereby increasing the heat input into the casting.

Buttering layer

The purpose of this is to deposit beads of weld along the edges prepared for welding. This completely blanks off the casting from the weld metal which is to be subsequently deposited. Buttering is carried out to advantage with a nickel alloy electrode, ensuring that the fusion zone is soft and ductile, has light penetration and is free from hard carbides. (See Fig. 47(a).) For economic reasons,

87

the subsequent filling-in runs of weld may be made by using soft iron electrodes. The soft iron deposit is not affected by the carbon in the casting because this has been blanked off by the previously deposited buttering runs of weld. Better welds are made on cast-irons by using this procedure, which is also the most economical. (See Fig. 47(*b*).)

(a)

(b)

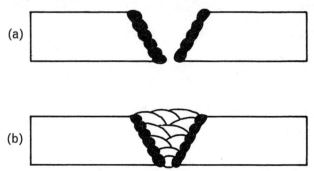

Fig. 47. (a) *Buttering of cast-iron.* (b) *Completed joint.*

Skip welding and peening

When castings are arc welded without a preheat, the main problems are reducing contraction stresses, minimizing hard zones, using a filler material which is capable of taking all contraction stresses and keeping the casting cool. The skip welding procedure is effective in reducing the heat input and balancing the effects of the contraction stresses. When skip welding, a casting is welded intermittently, each run of weld bead not exceeding $1\frac{1}{2}$ in in length. Immediately after each bead is deposited it should be lightly peened. The purpose of this is to expand the shrinking weld deposit and, in doing so, reduce the contraction stresses. (See Fig. 48.)

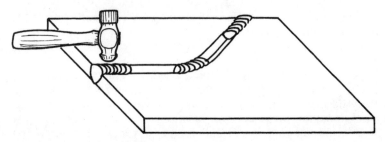

Fig. 48. *Skip welding and peening.*

The short runs of weld beads have the effect of reducing the heat input and thus preventing the formation of hard zones. In welding practice, the casting should be kept cool enough to allow the welder to place his hand upon the casting, about three inches away from any welded area; if this is not possible the casting is too hot.

Unweldable grey cast-iron

On rare occasions a casting may require repairing and it may defy any attempt that may be made to weld it. The reason for this is that the chemical condition of the material does not permit fusion with the weld metal to take place. These cast-irons may be regarded as burnt cast-irons. Cast-iron which has been frequently heated to a higher than normal temperature undergoes internal oxidation, which can be remarkably intense in certain cases. Atmospheric oxygen can penetrate grey cast-iron which is in the solid state by travelling along the interconnecting flakes of graphite in the casting. Thus oxides are produced throughout.

It is perhaps easier to understand if consideration is given to applying an oxy-acetylene blowpipe to the surface of a piece of burnt grey cast-iron. The material refuses to melt and gives off a dazzling light, similar in appearance to that produced when an oxy-acetylene flame is directed on to a fire brick. Even the presence of silicon in the filler rod cannot overcome this condition, which is a result of oxidation of the casting. These unweldable cast-irons are very rare, but it is as well to know of their existence and to be prepared for failures in the actual welding operation.

Malleable cast-iron

There are two different methods used to produce malleable cast-iron, an English and an American method.

English method

Malleable cast-iron may be obtained by decarburizing the surface of an ordinary white cast-iron. To do this the castings are packed into boxes containing iron turnings, hemanite and iron oxide. They are then placed into a furnace which is held at a temperature of 950°C for several days. By an excessively slow reaction, the oxide of iron gives up its oxygen to the carbon of the metal, forming

G

carbon dioxide. The result of this is that the surface of the casting is decarburized and it is now more like steel than cast-iron. Internally the carbide of iron has taken up the form of nodules. It should be understood that the reaction commences at the surface of the casting and spreads little by little to the interior of the mass.

Castings treated in this way are known as *whiteheart malleable cast-irons*.

American method

The American way of producing malleable castings necessitates the packing of the castings in a non-oxidizing gas atmosphere at a temperature of 850°C for several days. The carbon is not lost from the castings, but remains in the form of flake aggregate graphite nodules.

The appearance of the fracture in this material is black, hence the name *blackheart malleable iron castings*. This material is suitable for the casting of relatively small sections only.

The purpose of producing malleable cast-iron is to make a stronger and more ductile material. The entire process is relatively inexpensive.

The joining of malleable cast-iron

If the welding of malleable cast-iron is to be considered, thought must be given to what would happen if the malleable iron were taken to a high temperature and cooled either slowly or rapidly. Malleable iron reverts to white cast-iron if it is heated above the transformation range. The graphite, which separated during the long period it was held at during manufacture, dissolves and then reproduces brittle cementite on slow cooling to room temperature.

If a flux-coated electrode is applied to the surface and an arc is struck, regardless of what composition the electrode may be, there will be a rapid loss of heat from the weld area, and a quench effect upon the weld metal and the area adjacent to the weld. The result of this is to return the malleable cast-iron to white cast-iron in the zone adjacent to the weld, accompanied by the undesirable properties of hardness and brittleness.

It is possible to keep the weld deposit ductile, strong and un-affected by welding heat by using an electrode with a nickel content of 55% and the remainder iron. However, the behaviour of the hard brittle zone is unpredictable. If a satisfactory joint is to be made that will not fail when put into service, a full anneal of the heat-affected zone must be carried out. This requires the same heat treatment that was carried out to produce the malleable iron. The oxy-acetylene method of welding presents difficulties if fusion welding is attempted. In the case of a whiteheart casting, its surface layer, being decarburized, is similar to steel, but beneath it lies a core of cast-iron. Welding becomes difficult on account of the entirely different nature of the material between the surface and the core, i.e., different melting points, whether the oxides are fusable, etc.

In the case of a blackheart casting, the difficulties arise from cracking along the line of fusion, and this cannot be successfully corrected. Both types of malleable cast-iron may be successfully joined by bronzewelding. This is carried out at a much lower temperature than is required for fusion welding and, as a result, it does not destroy or seriously affect the malleable properties of the castings.

Bronzewelding of cast-iron

The term 'bronzewelding' would be more correct if it was called 'brazewelding', the filler alloy being of copper and zinc, or copper, zinc and a small amount of nickel. This method of joining cast-iron is equally applicable to all types of cast-iron, although the joining of white cast-iron should be regarded as a doubtful proposition. Bronzewelding may be regarded as the preferred method of joining malleable castings.

The parts to be joined are bevelled to an included angle of 90°, and care should be taken to see that the faces to be joined are thoroughly cleaned. A suitable flux should be used to remove any oxides which may form during bronzewelding. Use is made of the oxy-acetylene flame to heat locally the casting at the joint. It is important that the casting is only heated to the correct temperature, this being between 800° and 900°C, so that it is *not* melted. When at the correct temperature, the filler alloy should be applied. This will then flow evenly over the prepared surface, care being taken not to overheat the casting, thereby forming oxides on the surface and completely ruining any attempt at bonding.

At the surface, where bonding is taking place, the liquid filler alloy of copper, zinc and nickel forms a solid solution with the existing solid iron atoms when the temperature drops. The condition of this bonding zone is neither like cast-iron nor like the bronze, but should be considered as an intermediate zone because of the diffusion of the atoms having taken place. The solution, so formed, will be a substitutional solid type; its properties are quite favourable and it can be regarded as strong and ductile.

If this bonding is done correctly, it is only necessary to build up the joint to the correct height for success. Remembering that atoms are very small indeed, this in turn will account for the bonding zone being very small, but nevertheless, it is the most important part of the joint. (See Fig. 49.)

Fig. 49. Bronzewelded joint. The size of the bonding zone is deliberately exaggerated.

Spheroidal graphite cast-iron

This type of cast-iron is also known as nodular graphite cast-iron, nodular cast-iron, ductile iron and S.G. iron. This is a modern engineering material, distinguished by its high strength, toughness and ductility, combined with excellent casting properties and good machinability. It has a high modulus of elasticity and good resistance to corrosion and wear.

For many purposes, S.G. iron is replacing such established materials as grey cast-iron, malleable iron, non-ferrous alloys and cast or forged steel, with advantage and economy.

When grey cast-iron is considered, the graphite may be found in the form of elongated flakes, as seen in Fig. 50. Many of these flakes are interconnecting and create discontinuities, with the consequence that they act as stress raisers throughout the material, reducing the strength and ductility of the casting. A much stronger and more ductile material is obtained by making the graphite take up a spheroidal shape rather than existing as elongated flakes. To do this, a suitable cast-iron is chosen, with a carbon content on the high side of the range, and a silicon content from $2\frac{1}{2}\%$ to 3%. This means that the spheroidization process may be applied to any type of cast-iron which is capable of producing flakes of graphite when slow cooled.

During casting operations, if a small percentage of magnesium is added whilst the cast-iron is liquid, it will be retained in the iron and will ensure that the graphite takes up spheroidal form. A typical spheroidal graphite cast-iron is shown in Fig. 51. The graphite is not interconnected when in the form of spheroids, therefore there are no stress raisers. The material now has better mechanical properties which may be further improved by correct heat treatment.

Fig. 50. Grey cast-iron × 100.
Fig. 51. Spheroidal graphite iron. Pearlitic ('as cast') × 100.
Fig. 52. Spheroidal graphite iron. Ferritic (annealed) × 100.

Pearlitic spheroidal graphite iron

The 'as cast' structure shows that an appreciable amount of the carbon is present in the combined form, and the structure consists of intimately mixed ferrite and cementite with spheroids of graphite. (See Fig. 51.) This gives a hard, strong iron with about twice the strength of ordinary grey cast-iron.

Ferritic spheroidal graphite iron

When pearlitic spheroidal graphite iron is annealed, the structure becomes ferritic. This is because all the combined carbon is broken down and distributed as graphite spheroids. The background of the structure is ferrite. (See Fig. 52.) In this condition the iron shows maximum toughness and ductility, with increased elongation.

Welding and brazing of spheroidal graphite iron

Spheroidal graphite iron can be readily welded to itself, to mild steel and to many other materials by the manual metal-arc welding process. Electrodes, having core wires containing 55% nickel with iron alloys, give a deposit which is ductile and the results are very satisfactory. The joint matches the strength of the spheroidal graphite iron and the deposit is readily machinable. Adjacent to the weld, however, there may be a narrow hardened zone. In many applications this zone is not significant, but if necessary, it can be eliminated by a full annealing treatment after welding, or its effect can be greatly reduced by tempering at a temperature between 550° and 700°C. The ductility of spheroidal graphite iron makes it more readily weldable than ordinary flake graphite cast-irons, and there is therefore no need to preheat the ferritic grade. When welding the pearlitic spheroidal graphite irons, a moderate preheat of 200°C should be carried out.

In the oxy-acetylene welding of spheroidal graphite iron, the techniques usually employed for flake graphite cast-irons are suitable. A cerium-bearing, spheroidal graphite iron filler rod is used to provide weld deposits containing graphite in the spheroidal form.

Spheroidal graphite iron can be bronzewelded and brazed by using the conventional techniques. However, when such a joint is tested, the strength of the spheroidal graphite iron is such that failure is likely to take place at the bonding junction, not in the parent material, as so often occurs with the weaker flake graphite irons. It should be realized that the strength of a bronzewelded or brazed joint will be considerably less than the strength of the spheroidal graphite iron.

Unfamiliar terms used in this chapter

Spheroidal: A body which is globular shaped, i.e., nodular graphite.

Graphite: A form of carbon found in grey cast-iron or spheroidal cast-iron.

Flake graphite: In grey cast-iron, graphite is found in the form of flakes.

Volatilization: The loss of an alloy by the change of a material to a vapour state.

Decarburizing: Removing carbon from a material.

Nodule: See spheroidal.

Bonding: Joining or holding together.

Hardsurfacing: Depositing a hardened layer on a softer base material.

Stress raisers: Points where stresses may accumulate.

Chapter review questions

What forms may carbon take in cast-irons?

Which cast-irons are satisfactory for welding?

What are the properties of spheroidal graphite cast-iron?

What factors help to produce grey cast-iron after welding?

Why may cracking take place after welding grey cast-iron?

What steps may be taken to cut down cracking possibilities when welding grey cast-iron?

Which elements may be lost during fusion welding by the oxy-acetylene welding process?

What is the advantage of using a buttering technique?

When may a grey cast-iron be considered unweldable?

How should malleable cast-iron be joined?

Why is peening carried out immediately after welding?

Why are nickel alloy electrodes considered very satisfactory?

CHAPTER 9

Principles involved when Joining Dissimilar Materials

For many applications it is necessary to join together dissimilar materials. Quite often a welder who is engaged upon maintenance repair work will be called upon to join together materials such as mild steel and cast-iron. He will already know that there is a great amount of difference in the behaviour of the two individual materials and he would never attempt to melt them together by the oxyacetylene flame. These may be regarded as the materials that the welder is most likely to be asked to join together. It would be very unwise to consider joining a brittle material such as cast-iron to a ductile material like mild steel without taking special precautions. Because of the nature of the deposited filler metal, a mild steel or a cast-iron filler wire would be unsuitable. Therefore careful thought is necessary in choosing the correct type of filler material. Engineers often require the joining of copper to brass, steels or nickel. In fact there seems to be no end to the number of permutations available for the combinations of dissimilar materials which can be joined together.

Difficulties encountered when joining dissimilar materials

Expansion
   The rates of expansion of the two materials to be joined should be known. If the expansion rates are found to be quite similar to each other, then there is no special difficulty during joining. Should there be considerable difference between the expansion rates, then this may possibly be overcome by applying a higher preheat to the material with the lowest rate of expansion.

96

Conduction

It is necessary to consider both thermal and electrical conductivity.

Thermal conductivity. In order to overcome heat losses during welding, it may be necessary to apply a greater amount of heat to the material which has the best thermal conductivity. Preheating one plate more than the other is one way of doing this, whilst during actual welding the source of heat may be applied to concentrate more on the material which is the better conductor.

Electrical conductivity. All metallic materials conduct electricity throughout their length. Some pass current through their length relatively easily, such as copper, whilst others offer some resistance to current flow, e.g., stainless steel. The resistance of a material to current flow may be used to increase a small area of the material to a high temperature suitable for resistance welding. With dissimilar materials, the different resistances to current flow will result in unequal heating of the parts to be welded by resistance spot welding. Dissimilar materials may be joined, however, there being several available means of balancing the unequal heating when spot welding.

Oxidation

An oxide layer always prevents the joining of any two materials together, whether they are of similar or dissimilar composition. Precleaning, such as pickling, degreasing or wire brushing should be carried out before welding. The troublesome effects of oxidation may be prevented, provided that suitable protection is provided against atmospheric contamination.

Dilution of weld deposit

When a root run of weld is deposited between two edges to be joined, dilution of the weld material with the base material may be as high as 45%. If two mild steel plates are welded together, this dilution is not serious because there is no detrimental effect on the properties of the material. However, if dissimilar materials are considered, then the effects of dilution may be disastrous. When fusion welding dissimilar materials, dilution can be controlled by buttering one or both surfaces to be joined. When buttering, metal is deposited along the fusion face of the material. By doing this, the dilution with the base material is reduced to below 20%, which for many applications is acceptable. The object in cutting down on dilution is to prevent the formation of a hard, brittle constituent within the structure, thereby increasing the risk of cracking.

The desired aim when welding is to ensure that there is a minimum loss of ductility in the heat-affected zone and the weld metal.

Methods of joining

There are several ways of joining dissimilar materials together. The most widely used industrially are bronzewelding, brazing, flash butt welding, upset butt welding, spot welding, manual metal-arc welding, metal inert gas welding, friction welding and soldering.

Examples of joining dissimilar materials

It has been mentioned earlier that mild steel is often required to be joined to cast-iron. Bearing in mind the differences in the properties and the individual welding difficulties, this type of joint may now be considered.

Mild steel to cast-iron

The properties of these materials are the first consideration. Mild steel contains approximately 0·3% carbon and is ductile. Grey cast-iron contains between 2% and 4½% of carbon and is inherently brittle. Admittedly the expansion and contraction of the two materials are much the same but, because of brittleness, cast-iron may crack during contracting in the cooling stage after welding and a preheat is necessary to assist in controlling the contraction stresses. In order to keep the carbon out of the weld area the cast-iron is buttered, by depositing a nickel alloy material as shown in Fig. 53. Nickel does not form hard brittle carbides in the weld junction, so it can be deposited by the manual metal-arc process. After buttering, the remainder of the joint may be filled up with a soft iron deposit. This type of joint is ideal for many applications.

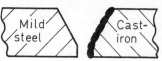

Fig. 53. Buttering layer deposited on cast-iron.

Bronzewelding may also be used for joining these two materials. During joining, the materials are not melted and because of this there is no chance of dangerous carbides being formed in the bronzeweld deposit. It should be remembered that, after welding, the joint should be reheated and cooled slowly. This assists in the prevention of cracking in the cast-iron through uneven contraction stresses.

Austenitic stainless steel to mild steel

This combination is often welded together. Dilution of the base material with the deposited metal is still a problem. In order to cut down on dilution, buttering of the mild steel by the manual metal-arc welding process is recommended. A suitable type of electrode for this buttering layer would contain 25% chromium and 20% nickel, thus ensuring that the fusion zone is austenitic, and therefore ductile. It will be noticed that the buttering layer is deposited with an austenitic electrode containing a high percentage of nickel. If this were not so, then a hard, brittle martensitic deposit which is prone to cracking would result. Should the nickel in the weld junction be less than 6%, caused by dilution, the hard martensite forms. The rest of the joint may be completed by using a more conventional 18% chromium, 8% nickel type of electrode. (See Fig. 54 and Fig. 55.)

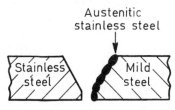

Fig. 54. Use of austenitic material for buttering mild steel.

Fig. 55. Completed joint.

Austenitic stainless steel to other materials

In the case of joining austenitic stainless steel to mild steels, the heat-affected zone, where the melting point is not reached, is ignored because of its ductile properties. When joining to hardenable steels, in addition to the problem of dilution, there is the risk of cracking occurring in the heat-affected zone as well. Again the manual metal-arc process is favoured, but, because of the localized high heat input, rapid cooling tends to cause a quench effect, forming martensite. To overcome this drastic result, preheating to a suitable temperature, i.e., 250°C, is necessary. To control the effects of dilution, a 25% chromium, 20% nickel electrode should be used to deposit a buttering layer on the hardenable steel. After the first run of weld, the dilution is greatly reduced, this means that the electrode to be used for depositing subsequent layers may now contain a smaller percentage of chromium and nickel. After

Welding Science and Metallurgy

welding, a postheat should be applied, following the treatment which is necessary for the hardenable steel base material.

Small sections of suitable shape, i.e., round bars, may be joined by the friction welding process.

Copper to nickel

Copper and nickel are capable of existing in the form of a solid solution over their entire range of combination percentages. It is well known that copper has excellent conduction properties which tend to make welding difficulties. Provided that the conduction of heat from the weld through the copper can be dealt with effectively, then, because copper and nickel are able to form a solid solution, joining will present little difficulty. Preheating of the copper to a larger amount than the nickel is necessary in order to bring both fusion faces to their melting temperature. Monel metal may be used as the filler material and this may be deposited by manual metal-arc, metal inert gas, or the oxy-acetylene process. Bronze-welding and brazing may also be used for joining these metals.

Copper to steels, stainless steels

If it is required to join copper to steel, then this is best carried out by the manual metal-arc process or by bronzewelding. When manual metal-arc welding, the first important factor to consider is heat losses by thermal conduction through the copper; this may be controlled by adequate preheating. A buttering layer of nickel may then be applied to the face of the copper to be welded. This

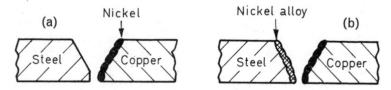

Fig. 56. (a) *Copper buttered with nickel.*
  (b) *Steel buttered with nickel alloy.*
  (c) *Completed joint made with monel deposit.*

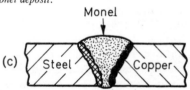

acts as a barrier to heat losses into the copper when depositing subsequent layers of weld metal. (See Fig. 56(*a*).) A buttering layer of nickel iron alloy should be deposited on the steel side of the joint. (See Fig. 56(*b*).) The remainder of the joint may now be filled up by depositing layers of weld metal using a monel electrode. (See Fig. 56(*c*).) It will be noted that the monel deposit is not made directly on to the steel; the reason for this being that a brittle material may be formed, resulting in possible cracking.

The other popular method of joining copper to steel is by bronze-welding. This presents little difficulty as both materials are capable of producing the necessary bonding layer essential to the success of this technique.

When copper has to be joined to austenitic stainless steel, the surface of the copper is again buttered with nickel. It is not necessary to butter the surface of the austenitic stainless steel. The reason for this is that the remainder of the joint may be completed by using an electrode of the same composition as the austenitic stainless steel itself. Due to the high thermal conductivity of the copper, preheating should be carried out and the heat level maintained during welding.

Clad steels

Nickel, monel, inconel or stainless steel are often used for corrosion-resisting conditions. Unfortunately, these materials are expensive, so, to cut costs, a thin layer of any of these materials is applied to a thicker steel section. By this action, the costs are effectively reduced and a suitable corrosion-resisting surface is obtained. On reaching the welding workshop, the clad material is in its final state, with its corrosion-resisting surface being about 10% to 20% of the total thickness. (See Fig. 57.)

Fig. 57. Clad steel.

The metallurgical problems that can arise during welding are quite similar to those that occur when joining two dissimilar materials together, i.e., when welding a piece of mild steel to a piece of austenitic stainless steel. Should dilution between the clad material and the steel base take place whilst welding, there is a possibility of martensite, hardness and possible cracking occurring. Usually, with the type of joint that is employed, safe policies may

101

be put into operation to ensure that dilution does not take place. 'Never, never penetrate the cladding' is now a time-honoured cry when the welding of dissimilar materials in the form of clad steels is carried out. It is not to be ignored either. If for any reason it is, then the consequences which follow are usually disastrous, leading to failure of the welded joint. For practical welding purposes the manual metal-arc process is preferred, although sound joints can be made with the metal inert gas process. With either process the steel side *must* be welded first, using an electrode which is of similar composition to the steel base metal itself. The preparation of the joint should be as shown in Fig. 58. It will be seen that there is a root face on the steel side of the cladding. This is to ensure that excessive penetration leading to dilution does not take place.

Welding of austenitic, stainless steel clad steels

Consider welding the steel side on an austenitic, stainles steel clad material. If penetration is excessive and only about 3% chromium is diluted with the weld metal, the cooling rate will be fast enough to form martensite. This results in hardness, brittleness and possible cracking.

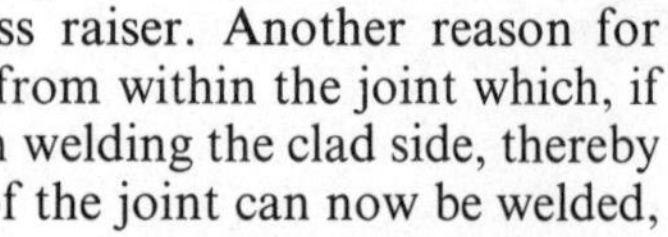

After completely filling the steel side of the joint, it is necessary to turn the clad plates over and chip out until clean fused steel is showing.

Fig. 58. Clad steel joint prepared for welding.

The purpose of backchipping is to remove unfused metal which may otherwise act as a stress raiser. Another reason for backchipping is to remove any slag from within the joint which, if not removed, may turn to a gas when welding the clad side, thereby resulting in porosity. The clad side of the joint can now be welded, using an electrode of suitable alloy composition for the purpose. During welding, penetration of the steel deposit takes place; this is not detrimental because the iron pick-up is not sufficient to alter the soft ductile structure òf the deposited alloy. It should be remembered that in this example the clad material is austenitic stainless steel which contains at least 18% chromium and 8% nickel. To allow for any dilution with the steel base, the first run of weld on the clad side of the joint should be made with a 25% chromium, 20% nickel electrode. The remainder of the joint may then be completed with an electrode of the same composition as austenitic stainless cladding itself. (See Fig. 59.)

102

The manufacturer of the austenitic stainless clad material should
ensure that the cladding contains
suitable stabilizing elements to
prevent intergranular corrosion.
In order to prevent intergranular
corrosion within the weld deposit,
the electrodes should be stabilized
with niobium.

Fig. 59. A completed weld between two pieces of clad steel.

Welding of nickel clad steels
The welding of nickel clad steels presents problems related to
iron pick-up which cannot entirely be prevented. The purpose of
using nickel as the cladding metal is for certain corrosion-resisting
applications. It is essential that the filler material should contain
chromium to compensate for the iron pick-up, which decreases the
corrosion resistance. However, iron dilution will still take place,
but it can be controlled by removing half of the first bead of weld.
The weld is then completed by depositing straight beads.

Welding of chromium clad steels
The undesirable brittle effects produced by dilution can be
avoided by welding the clad side of the joint with an austenitic
stainless steel electrode. On thicker sections of this type of material,
i.e., ⅜ in thickness, preheating and postheating is necessary to
relieve internal stresses and remove any brittle structure which may
be present in the heat-affected zone. The preheat temperature is
about 200°C, and after welding, postheating must be carried out at
700° to 750°C.

Welding of thin clad sections
Some of the clad steels which are welded have a very thin clad
surface. To prevent burn through and dilution when welding, it is
advisable to weld the entire joint with an alloy type of electrode.
The composition of the alloy electrode should preferably be of a
higher alloy content than the clad surface being welded. This is to
prevent the dilution, which takes place from impairing the corrosion resistance of the joint.
It is particularly important to realize that weld defects, such as
undercut, weld craters and porosity, may leave the steel without the
protection the clad provides against corrosion. The reason for this
is that the depth of the defect may be greater than the thickness of
the clad material.

Welding processes suitable for clad steel

The manual metal-arc welding process is considered very suitable for the welding of clad steels. The manual control of the penetration depth is a great asset in avoiding contamination of the weld deposit. Another suitable process is the metal inert gas welding process. The shielding gas is preferably argon.

Unfamiliar terms used in this chapter

Dilution: To mix liquid metal with liquid base metal.
Monel: An alloy of nickel and copper.
Backchipping: To remove metal from the root of a weld to facilitate depositing a run of weld on the back side.

Chapter review questions

Give three problems associated with the joining of dissimilar materials.
What are the advantages of buttering?
How may copper be joined to steel?
What are clad steels?
Name four precautions to be taken when welding clad steels.
Why is the steel side welded first when welding clad steels?
What is the object of cladding?
How may dilution be avoided?
What would the result be if dilution of the clad took place when welding the steel side?
How would thin clad steels be joined?
Why should weld defects especially be avoided when welding clad steels?
What physical properties should be considered when joining dissimilar materials?
When welding austenitic 18/8 stainless clad steels, what electrode would be suitable for the root run of weld on the clad side?

CHAPTER 10

Considerations when Hardsurfacing

Hardsurfacing of materials is considered to be a modern development. It may seem strange to relate that hardsurfacing has been carried out for a large number of years. It is probable that the first instance of hardsurfacing being carried out was the hardening of swords, axes and arrows in ancient times. Man learned that the blade of a steel sword could be hardened by holding it within the embers of a wood fire. This resulted in the outer surface of the blade being increased to a temperature where it was able to absorb carbon which was released from the burning wood. Therefore the outer surface of the blade was carburized to provide resistance to wear, whilst the interior retained good ductility. With the passing of time, further thought was given to the hardening of farm implements for the purpose of cultivation. Old time blacksmiths heated implements and hammered small chippings of cast-iron into their surfaces. After quenching, followed by a form of tempering, the life of the farm implement was considerably extended. At the beginning of the twentieth century the oxy-acetylene flame was introduced. It is common knowledge today that the use of a carburizing flame will allow carbon to pass into the surface edges of steel if the temperature is sufficiently high. The result of this is an increase in the surface hardness. Further developments have made hardsurfacing a very important and widely used industrial application. Use is still made of the carburizing of mild steel for hardsurfacing, in fact the case hardening principle relies upon carbon being induced into steel whilst in the solid state. Welders will realize that this may be applied to the tip of a chipping hammer, but, through continually chipping hot slag and welds, the tip soon becomes soft again. If the tip is surfaced with a suitable hard, heat-resisting material, then it will last a considerable length of time without having to be reground for further use.

In the last twenty years, the use of hardsurface deposits has been continually expanding in many industries. The application of a hardened surface layer to the surface of a softer base material is of great importance when reclaiming worn parts. Another application is to extend the life of new components by depositing a hardsurface layer to the area which may be subjected to any form of wear or corrosion.

Some of the many uses of hardsurfacing are to be found in excavating equipment, mining and oil drill equipment, internal combustion valves and seats, machine parts and shear blades for cutting various metallic materials. There is no single type of hardsurface deposit that will cover all the different applications. There is, however, considerable overlapping of their uses.

Types of deposit

In order to understand which type of a deposit to use for a given application, it is necessary to understand what is meant by the term 'wear by abrasion' or 'wear by impact'.

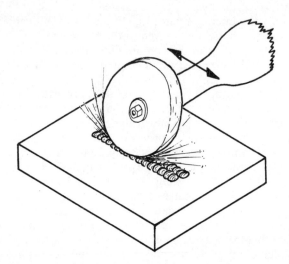

Fig. 60. Abrasive action of a carborundum wheel.

Abrasion

This is the continual action of one material scratching or grinding on the surface of another material. It may be in the form of metal to metal, or metal to sand or gravel, etc. Fig. 60 shows an example of abrasion being put to a desirable use by an abrasive carborundum wheel levelling the surface of a completed weld. A macro specimen is prepared by the abrasive action of a file, followed by the finer abrasive action of an emery cloth. The continuous action of gear wheels contacting each other causes wear by abrasion to take place. Use may be made of depositing wear-resisting materials on the individual teeth.

Impact

The continuous pounding action of a hammer on the surface of a material illustrates impact. (See Fig. 61.) Impact wear occurs when the impact-producing material presses more or less in a downwards direction upon the surface which wears. A fine example of wear by impact occurs when railway rolling stock passes over points; because of this action it is necessary to surface the points with a suitable impact-resisting material.

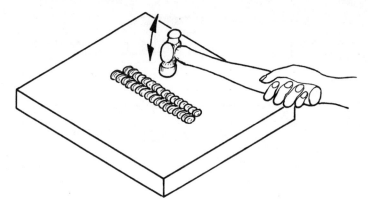

Fig. 61. Wear by impact.

It should be pointed out immediately that a material that has good resistance to wear by abrasion may have low resistance to wear by impact. The different groups of hardsurfacing materials are shown in Fig. 62.

Tungsten carbide group

It will be seen from Fig. 62 that the materials that have the highest resistance to abrasion are the tungsten carbides. In the cast-tungsten carbide deposit it is usual to find the crystals of hard tungsten carbide embedded in a background of tough tungsten and carbon alloy. The alloy melts at over 2 500°C. Therefore it can be stated that a tungsten carbide alloy may be deposited by the manual metal-arc, carbon-arc or oxy-acetylene process, without being

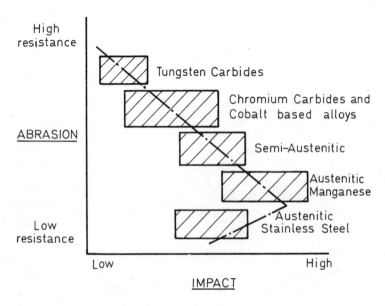

Fig. 62. Relationship between abrasion and impact resistance of various hardsurfacing materials.

melted. For hardsurfacing purposes, tungsten carbide may be obtained in the form of granules which are contained in a steel tube. These carbide-containing tubes may be deposited by the electric arc or flame. It is usual to join the carbide particles on to the surface of a component by immersing them in a molten steel puddle obtained from the enveloping steel tube. The steel tends to pick up a small amount of carbon and tungsten when depositing, therefore the final wear-resisting product is composed of three constituents. These are:

   (*a*) The granular tungsten carbide particles.

   (*b*) The hard tungsten alloy that holds the carbide crystals.

   (*c*) The strong iron alloy holding the cast-carbide granules (iron alloy obtained from the steel tube).

The steel tubes containing the tungsten carbide granules are readily applied by the manual metal-arc, carbon-arc or oxy-acetylene process.

A special application of the oxy-acetylene process is the use of tungsten carbide in the form of slugs. The slugs are inserted $\frac{3}{8}$ in apart into the molten surface of the tool, and allowed to sink one-third of their diameter. This is followed by depositing over the entire surface containing the slugs a layer of steel for the purpose of holding them in position. To prevent wear of this binding layer of steel, a deposit of fine mesh carbide particles is applied from the steel tube which contains the carbides. In service, the material wears between the inserts, leaving them standing proud as individual cutting teeth. A partially worn cutting tool is shown in Fig. 63.

Fig. 63. Partially worn tool showing carbide inserts.

It is possible to obtain tungsten carbide particles graded according to their size or coarseness. The largest granules of tungsten carbide give the greatest abrasive-resisting surface. This type of deposit is therefore suitable for surfacing drilling bits, digger teeth, tractor treads, coal picks, etc.

When depositing the smaller size granules, some of them will dissolve in the hard tungsten alloy that holds them. Whilst this type of deposit is still extremely hard, the resistance to abrasion is reduced; but the deposit is tougher. These find uses on surfacing plough shares, sprockets, hammers, etc.

Chromium carbide group

Chromium carbide plays an important part in materials used for hardsurfacing. One type of deposit contains ferrite as a base

Welding Science and Metallurgy

material with up to 30% chromium and a maximum of 5% carbon. Other elements such as manganese and molybdenum may be present in small amounts. It is the combination of chromium and carbon that forms the very hard chromium carbide. These alloys are abrasion resistant and they are tougher than the tungsten carbide group. Their resistance to corrosion is quite high and some of the higher chromium alloys retain high hardness at elevated temperatures.

The chromium carbide type of deposit may be obtained from either covered electrodes or in the form of powder. When electrodes are used, they are applied for surfacing materials in excess of ⅛ in thick. For depositing upon thinner materials, the chromium carbide powder is suitable. This may be deposited by using the carbon-arc or the oxy-acetylene process. A typical example of chromium carbide powder being used is in the placing of a hard deposit along the end of a spade or shovel, there being no risk of burning through the material, even though it may be quite thin.

Low alloy steel group

The general effect of the addition of alloying elements to steel is to increase the hardness when rapidly cooled, even though their total may come to under 5%. It is a safe rule when dealing with low alloy steels to treat them in a similar manner to high carbon steels. This applies to steels containing up to the alloy content shown below, which are normally not air hardening.

Carbon	0·45% max
Chromium	1·5% max
Molybdenum	1·0% max
Nickel	3·5% max

With this type of deposit, when hardsurfacing is carried out with the manual metal-arc, a minimum preheat of 250°C should be carried out if the carbon content is low (below 0·25%) and if the carbon content is higher, then preheat 300°C.

If depositing is carried out by oxy-acetylene, the heat supplied by the blowpipe is adequate for preheat with small articles which are made from mild steel. After hardsurfacing, the entire article is brought up to an even, dull red heat, followed by slow cooling. When dealing with large articles, care must be taken to allow adequate time for the preheating temperature to be attained throughout the article, a safe rule being to allow a minimum half-hour per inch thickness. This is because severe stresses will be set up in thick sections of steel by rapid heating and expansion of the surface layers before the centre has become hot. The article may be machined to finished size whilst in the softest condition and then reheated and hardened.

Two other popular compositions of low alloy steels for hardsurfacing are as follows:

(a) 0·1% carbon, 1·0% chromium, 0·5% molybdenum, remainder iron.

(b) 0·25% carbon, 2·5% chromium, 1·0% molybdenum, remainder iron.

These are deposited by the manual metal-arc method. The carbon and alloy content of these electrodes ensure a controlled degree of hardening during cooling of the deposit, but the low alloy electrodes are sufficiently soft after deposition, if a suitable preheat and procedure are used. Because of this the deposit may be machined.

On the other hand, if rapid cooling takes place, the deposit practically has its full hardness. The hardness attained depends primarily on the composition, but it can be varied by the welding procedure. A short bead with a small gauge electrode on a cold base material is harder than a heavy bead deposited with a heavy gauge electrode on a preheated base material. The cooling rate of the latter is slower, and the quenching effect less because of the larger heat input. This type of deposit is used for repairing shafts, rails, keyways, gear teeth, etc. Owing to the low cost of a new component, the cost of a hardsurfacing electrode must be competitive. These types of electrodes are relatively cheap. The deposit is amalgamated by fusion with the base material and penetration of the hard facing alloy is from $\frac{1}{32}$ in to $\frac{1}{16}$ in depending upon the type of electrode used. The deposit after machining is often heat treated to produce harder structures such as martensite.

Cobalt based group

This group is known all over the world as stellite. It is composed of a cobalt base which alone is a comparatively soft element, but when alloyed with tungsten, chromium, molybdenum and carbon produces a very hard wear resisting material. Various combinations of the alloying elements give different degrees of toughness and hardness. They have a high wear resistance at normal and elevated temperatures, and, in general, they are stainless and oxidation

111

resistant. Owing to their high hardness they cannot normally be rolled or drawn and they are cast in the form of rods. A thin hard surface layer is deposited to an average thickness of $\frac{3}{32}$ in. If a component is badly worn, it is built up by a softer steel type of filler material to within $\frac{3}{32}$ in of the required height. These alloys are used particularly where expensive preliminary machining of the surface and final grinding of the deposit are necessary. The purpose of this is to obtain the best possible life from the hard surfaced component. The high cost of the stellite rod is relatively unimportant. The method of deposition by the oxy-acetylene blowpipe is carried out by using a carburizing flame, the excess of acetylene being twice the length of the inner cone of a neutral flame. This adds carbon to the heated surface, causing the appearance of the surface layers to be 'sweating' when viewed through welding goggles. To obtain satisfactory results, the cobalt based alloy is not puddled or intermixed with the steel base metal. The filler rod is allowed to flow into the sweated surface producing a straight line bond, this also ensures a minimum amount of grain growth in the base material. The risk of cracking during cooling may be avoided by a suitable postheat of the component to dull red, followed by slow cooling. This may be done by covering the entire article in lime or mica.

The resistance to abrasion of the cobalt based alloys mainly depends upon the chromium and the carbon; the tungsten and molybdenum are added for hardness and strength at high temperatures. By the nature of their composition, the alloys are stainless and resistant to oxidation at high temperatures. The chromium content is usually at least 15% and the tungsten varies from 5% in the softest grades to 18% for the hardest grades. In the softer alloys of less than 8% tungsten, there are usually no free carbides. These softer alloys are used extensively for hard-facing internal combustion engine and steam valves, dies and shears for hot and cold work. The deposit may be machined with tungsten carbide tools.

With more than 8% tungsten, free carbide granules are formed, composed of chromium and tungsten carbides. Because of the formation of these carbides, the deposit is of high hardness and this composition gives the highest resistance to wear. However, the deposit is less tough and care is needed to prevent cracking when it is being deposited. This alloy is used for hard facing field implements, cams, crushers, scrapers and applications where little shock is encountered.

Cobalt based alloys may be deposited by the manual metal-arc process. Hard-facing electrodes should always be flux coated to reduce oxidation and loss of alloying materials, and to provide a smooth bead free from holes and flaws. Electrodes should be kept dry; if their coatings are damp and deposition is attempted, excessive porosity and metal losses due to spatter may result. When depositing, a short arc should be maintained to prevent losses of alloying elements when transferred across the arc. A minimum of penetration is necessary to keep down the amount of dilution of the alloy. Some dilution of the base material is inevitable, so it is usual to deposit at least two layers. This leaves the second layer relatively free from dilution with the base material, in fact with highly alloyed or non-ferrous electrodes, three layers may be necessary to attain optimum properties. (See Fig. 64.) If more than

three layers of stellite are deposited, then cracking between the deposit and the base material will occur. If a worn part requires more than three layers, then the lower runs of weld should be deposited with a low hydrogen type of coating applied on a mild steel type of electrode.

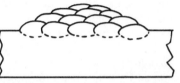

Fig. 64. Correct deposition of layers.

For industrial use, the cobalt based alloys are available in three grades, these being grade one, grade twelve and grade six.

Stellite Grade One
This is used when extreme abrasion resistance is required. However, its resistance to impact is quite low, the hardness number being Rockwell C 55.

Stellite Grade Twelve
This type of deposit is suitable for components subjected to both abrasion and impact wear. It is softer than grade one but may be regarded as tougher, the hardness number being Rockwell C 48.

Stellite Grade Six
This gives the highest resistance to impact and hot cracking. It may be regarded as the softest and toughest of the three grades, its hardness number being Rockwell C 40.

113

Semi-austenitic group

This is a very popular group of hardsurfacing materials. The amount of alloying elements contained in the deposits being less than 20%. The most popular types of electrodes contain 1% or 2% carbon and between 5% and 12% chromium, 4% nickel, the base metal being iron. The name 'semi-austenitic' may be difficult to understand. When being deposited by the manual metal-arc process, at the *instant* of solidification the structure is entirely austenite. On cooling to room temperature, the *speed* of cooling determines the amount of austenite which remains in the structure, if any. Therefore the final structure is predominantly austenite or a combination of martensite with a small amount of austenite. Because the rate of cooling can be controlled, the final structure can be varied to suit the particular application of the parts being surfaced, i.e., to give an austenitic deposit, if the parts are subject to wear by impact, or a predominantly martensitic structure, if the parts are subject to wear by abrasion.

Fast cooling

Fast cooling rates may be obtained with the manual metal-arc process by using small gauge electrodes, depositing short beads and single layers on a cold base material. The effect of fast cooling is to provide a final structure in the deposit which is predominantly austenitic, the cooling rate being so fast that the austenite deposit does not have time to change to any other structure. The presence of nickel tends to suppress the change. These deposits, containing a lot of austenite when cool, are very tough but not very hard when deposited. They are work hardenable and rely upon pounding to change a thin surface layer to martensite with its desirable hardness.

Slow cooling

A slower cooling rate may be obtained by the use of a larger gauge of electrode, by continuously welding or by using a preheated base metal. The result of this is that the cooling rate is less drastic, there being sufficient time for some of the austenite to transform. Because of the influence of the alloying elements, the transformation will not be to ferrite and pearlite, but to the hardest structure in steel, martensite, which is resistant to wear by abrasion.

Regardless of whether there is a fast or slow rate of cooling, both deposits work harden to the same value. If they are first annealed at high temperature, the semi-austenitic deposits are machinable with tungsten carbide tools. It is a special claim of this group that they are capable of withstanding wear by both impact and abrasion.

When depositing a semi-austenitic steel electrode, it should never be used as a first layer on a carbon steel. The reason for this is that a hard brittle layer would result. The first layer should be made with a suitable *fully* austenitic stainless steel electrode, which will form a ductile bond with the base material. After this, the semi-austenitic deposit may follow. The reasons for the use of a buffering layer of a fully austenitic deposit is discussed in detail, on page 117, when austenitic manganese steel is dealt with, the reason being precisely the same in each case.

Austenitic deposits

If a material is subjected to conditions which may cause it to wear by impact, then an austenitic type of deposit is suitable. The austenitic steels used for hardsurfacing can be split into two groups, austenitic manganese steels and austenitic stainless steels.

Austenitic manganese steel

Austenitic manganese steel contains around 10% to 14% manganese and 1% to 1·4% carbon, the remainder being iron. If this composition is heated to 1 000°C and quenched, the effect of the alloying elements is to delay the transformation of the austenite to the extent that it is retained at room temperature. Because of the quench, the carbides are kept in the austenite solid solution, and the final cooled structure looks like an austenitic stainless steel when viewed under the microscope. This is the standard treatment to which an austenitic manganese steel is subjected. (See Fig. 65.)

It is strange but true that the austenitic manganese steel, in this condition, is quite soft, being about 200 Brinell. However, when pounded, it rapidly work hardens. Austenitic manganese steel is not much more resistant to wear by abrasion than mild steel when considering the abrasive action

Fig. 65. As quenched austenitic manganese steel.

115

caused by fine sand. However, it scores heavily over other materials when subjected to wear by heavy impact such as rock pounding. The more it is hammered and battered, the harder and stronger it becomes on the surface, remaining, at the same time, soft and ductile beneath the surface. In the course of service, wear occurs, but as the work hardened layer is worn away, the exposed surface is hardened.

Now that the austenitic manganese steel has been discussed with regards to its properties after quenching, slow heating and slow cooling may be considered. If austenitic manganese steel is heated slowly to a temperature above 450°C, the manganese will combine with the carbon forming manganese carbide, which will be precipitated out of the austenitic solid solution to the grain boundaries. (See Fig. 66.) The effect of the precipitated carbides is general embrittlement of the alloy. By comparison with hardenable carbon steels, an austenitic manganese steel is made less ductile when heated and slow cooled. Whereas a hardenable carbon steel remains ductile.

Fig. 66. Austenitic manganese steel showing carbide precipitation.

Hardsurfacing with austenitic manganese steel

When considering a welding process for depositing this alloy, the manual metal-arc process is the most suitable. The reason for this is the localized heat and quick cooling of the deposit. The electrodes used contain other alloying elements in the form of nickel, molybdenum and copper, in addition to the iron, carbon and manganese. The purpose of these additional alloying elements is as follows:

Nickel. Nickel is added to maintain the yield strength and improve the elongation percentage. The amount added is usually about 3% to 4%. Its presence tends to stabilize the austenite, thus assisting in preventing the precipitation of carbides, which may take place during cooling. It must be understood that a quench from a high temperature is still necessary when the maximum toughness is required. However, the cooling rate need not be as drastic as is the case when nickel is not present.

Molybdenum. This element is added in very small quantities of between 0·2% and 0·6%. The effect of molybdenum is to give the deposit higher hardness after cold working.

Copper. The amount of copper added is 1%. The purpose is to prevent carbide precipitation and to cut down the risk of shrinkage cracks when cooling.

When welding, care should be taken to keep the job being welded and the deposited material as cool as possible. A safe rule is to keep the job being surfaced below a temperature of 200°C at a distance of three inches from the welded hardsurfacing deposit. If this rule is broken, then, at a temperature of 450°C or above, manganese carbides form, causing embrittlement in the weld and the heat-affected zone. After welding, it is possible to reheat the austenitic manganese steel to 1 000°C and quench to obtain the best properties. Often, however, this is neither practicable nor desirable. By a good code of operational procedure the work may be kept cool so that after-weld treatment is not necessary. This can be in the form of depositing short runs of weld, skip welding, etc.

Quite often, materials such as mild steel are required to be hardsurfaced with an austenitic manganese steel deposit. The contraction rate of the manganese steel is 50% more than that of a carbon steel. Obviously the different rates of contraction lead to shrinkage stresses. In the first layer, dilution of the deposit with the base material takes place. This produces an alloy which is low in manganese and very brittle. This, along with the shrinkage stresses, results in cracking. It should be realized that at least 10% manganese is required to produce an austenitic deposit. If this is reduced by dilution, then a martensitic structure is formed. To overcome the undesirable martensitic structure, it is advisable to deposit a layer of austenitic stainless steel on to the base material

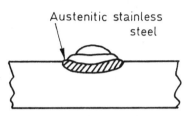

Austenitic stainless steel

Fig. 67. Austenitic stainless steel deposit providing an austenitic base to allow austenitic manganese steel (or semi-austenitic steel) to be deposited.

117

before the manganese steel is applied. When austenitic manganese steel is deposited upon austenitic stainless steel, the dilution creates no problems, the entire structure remaining austenitic and therefore ductile. (See Fig. 67.)

Austenitic stainless steel

Austenitic stainless steels are classified with hardsurfacing materials because they are only just one stage below manganese steels in their work hardening ability. They are able to withstand severe pounding without fracture, which makes them ideally suited when the component requires to be impact resistant in corrosive conditions.

This type of deposit is often used in the welding workshop when a worn component requires building up by welding. Very often, and perhaps without any knowledge of hardsurfacing, the welder concerned selects an austenitic stainless steel electrode to rebuild the worn part. Intuition tells him that this will increase the life of the component in service. Such electrodes should not be used indiscriminately, in spite of the successful results obtained. If corrosive conditions are encountered, then austenitic stainless steel is ideally suited for the application. A more economical method, when corrosive conditions are not serious, is to use a combination of austenitic stainless steel and austenitic manganese steel. Consider a mild or medium carbon steel which is to be hardsurfaced, or in fact any hardenable steel. If a deposit of austenitic manganese steel is deposited directly on to the surface of the steel, then dilution of manganese with the steel base produces a hard and brittle martensitic zone. This will result in cracking. To prevent this, if the first run of the deposit is made with an austenitic stainless steel, preferably of 25% chromium, 12% nickel, a soft, ductile austenitic deposit is produced. The buffering layer of austenitic stainless steel bonds well with the steel base material and also with the austenitic manganese steel. Thus a hard, brittle and martensitic layer is prevented. (See Fig. 67.) It is now possible to deposit the subsequent layers of austenitic manganese steel with satisfactory results.

Unfamiliar terms used in this chapter

Carbides: These are chemical combinations of a metal and carbon.
Granules: Small particles.

Straight line bond: This is penetration controlled to a regular depth.

Mica: A soft lime-like material in appearance, used to cover articles to allow slow cooling.

Buffering layer: A deposit placed on the surface of an article before a hard-facing layer is applied.

Chapter review questions

What is the purpose of hardsurfacing?

Name two ways in which wear may take place.

List the hard-facing materials used, in their correct order of abrasion resistance.

How may tungsten carbides be deposited?

State two different types of austenitic deposits.

Why is a maximum of three layers used when depositing hard-surfacing material?

If more than three layers are required to reclaim a worn surface, how may this be done?

What is the purpose of having different grades of stellite?

Which material may be subjected to battering and pounding more than others?

Give a use for surfacing with a martensitic deposit.

When using semi-austenitic deposits, how does the rate of cooling effect the final properties?

Describe two methods of depositing chromium carbides.

Why are alloying elements added to an austenitic manganese steel electrode?

Why is tungsten added to a cobalt based alloy?

CHAPTER 11

Aluminium, Magnesium and their Alloys

Aluminium

The quantity of aluminium used in industry is increasing continuously. Perhaps one of the main reasons for this is that the design engineer is no longer only restricted to the use of butt joints when considering a fabrication to be welded.

Aluminium is extracted from the ore bauxite. A large percentage of bauxite is in the form of aluminium oxide, known as alumina. Pure aluminium is produced by an electrical reduction process.

Pure aluminium

Aluminium, as used in industrial applications, may be up to 99·5% pure. The small amount of impurities left in the aluminium are iron and silicon, which are extremely difficult to remove completely when the aluminium is obtained from bauxite.

Mechanical properties

In the 'as cast' condition aluminium is a very weak metal. The tensile strength is as low as 5 tons per square inch. When aluminium is subjected to cold working, the grains break up and become elongated, the result of this is an increase in the hardness and the strength but this is accompanied by a reduction in the ductility. By subsequent annealing, the effects of cold working are removed. Annealing can be carried out by heating the aluminium uniformly to a temperature of about 350°C and allowing it to cool in air or by quenching in water. The effect of annealing is to increase the ductility. In the 'as cast' condition, the elongation percentage may be only 5%, but after annealing, this is increased to between

35% and 40%. Annealing lowers the tensile strength and hardness which resulted from cold working. The mechanical properties of aluminium may be improved by the addition of alloying elements and correct heat treatment.

Physical properties
Aluminium has a low melting point of 659°C. When heated to this point there is no colour change. Perhaps the most important property of aluminium is its extreme lightness—its density is only 2·7 compared with 7·8 for steel and 8·8 for copper. The specific heat and latent heat values of aluminium are quite high; it requires as much heat to melt one pound of aluminium as is required to melt one pound of mild steel. This is quite remarkable because there is almost 800°C difference between their respective melting points. It should also be remembered, however, that there is a greater volume of aluminium than in a similar weight of steel. Therefore a comparison of amperages when arc welding, or nozzle sizes when oxy-acetylene welding, cannot easily be made.

Aluminium is an excellent electrical and thermal conductor. Combined with steel strands, for added strength, aluminium is sometimes used for electrical conducting cables where good electrical conductivity is an essential factor.

The expansion rate of aluminium is considerably higher than mild steel for a given increase in temperature. Because of its high expansion and conduction values, preheating of thicker sections is usually carried out before welding. The preheat temperature should be a uniform 350°C.

Chemical properties
Aluminium forms a protective oxide film upon its surface. The effect of the film of oxide is to protect the pure aluminium from further oxidation. Aluminium oxide has a very high melting point of 2 050°C. Because of this, when welding is carried out the surface oxide cannot be removed by a melting action. In an attempt to achieve this the aluminium would melt at 659°C and the oxide would still be present in the solid form, resulting in the complete collapse of the molten aluminium beneath the surface oxide.

Aluminium alloys

Aluminium can be alloyed with other metals to obtain better properties. These include an improvement in the strength and hardness, the production of an alloy which is ductile throughout its cooling range, assistance in casting operations and an increase in the fluidity of the molten weld pool when welding. The aluminium alloys may be conveniently split into two groups, namely *heat-treatable* and *non heat-treatable* alloys.

Non heat-treatable alloys

These alloys contain additional elements such as silicon, magnesium, iron, nickel and manganese. In the annealed state this group of alloys is stronger and harder than pure aluminium. Each of the alloys may be further strengthened by subsequent cold working. They may be obtained in both the wrought and cast state, depending upon service requirements.

In the welding industry the aluminium silicon alloy has many applications. The beneficial characteristics of 5% to 15% silicon in aluminium include high fluidity, low shrinkage values on cooling and a ductile deposit throughout the range. During cooling after welding, cracking can result in certain alloys in the weld deposit because of brittleness. Therefore the aluminium silicon filler material is widely used because of its ductility. This enables it to withstand contraction stresses on cooling without cracking.

An important application of the 12% silicon aluminium alloy is its use as the filler rod when flame brazing pure aluminium.

The welding of pure aluminium and non heat-treatable alloys

The properties of pure aluminium and its non heat-treatable alloys show the difficulties that the welder has to overcome in welding aluminium. Regardless of the composition of the aluminium alloy, the first factor of importance is the removal of the surface oxide film. This may be done either before or during actual welding. Protection must also be provided against the oxide re-forming whilst welding is in progress. Oxide may be removed by the following methods:

(1) Mechanically. This entails filing or wire brushing the edges to be welded.
(2) Chemically. Before welding, the parts should be subjected to a pickling operation. This necessitates immersing the components to be welded in an acid solution at 60°C for half an hour. A suitable solution contains 20% sulphuric acid and

5% chromic acid in water. Pickling is particularly important when parts are to be welded by the M.I.G. and T.I.G. processes.

When aluminium or its alloys are welded with the oxy-acetylene welding process, a flux is used. The flux turns liquid at about 80°C below the melting point of the material. When molten, the flux dissolves the oxide film, leaving an oxide-free clean surface.

(3) Cathodic cleaning action. Advantage of this action is taken when aluminium is welded by the metal inert or tungsten inert gas welding processes. Consideration should be given to the metal, inert gas welding process. The welding supplies at the arc are d.c. with the consumable electrode connected to the positive terminal, the work therefore being connected to the negative terminal. The flow of electrons from the plates being welded breaks up the oxide film and disperses it from the weld area.

The same action takes place when welding with the tungsten inert gas process; the difference in this process is that the electrode is tungsten and a.c. electrical supplies are used. Every half-cycle the electrode changes polarity. When the electrode is positive, the cleaning action takes place. (See Fig. 68.)

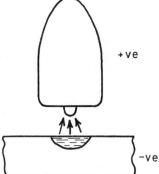

+ve

−ve

Fig. 68. Cathodic cleaning action when T.I.G. welding aluminium.

Preheating

To offset heat losses by conduction, to increase the speed of welding and to control the effects of expansion and contraction, a preheat is often necessary. The preheat temperature should be 350°C.

Filler material

When pure aluminium is to be welded a filler wire of similar composition is used. When aluminium alloys are welded, several matching type filler materials are available, the most popular being the types containing silicon as an alloying element.

Flux removal

The use of flux is essential with the oxy-acetylene welding process. It must be completely removed after welding, otherwise corrosion will take place. Flux removal can be carried out by scrubbing in hot soapy water, followed by a rinse in a 5% nitric acid, hot water solution. It can then be followed by rinsing in hot water.

Effects of welding upon cold-rolled non heat-treatable alloys

When two pieces of cold-rolled aluminium alloy are considered, the grains are elongated in the direction of rolling, as shown in Fig. 69(*a*). After welding, because of heat losses from the weld into the plates being welded, three distinct zones can be seen in the microstructure. (See Fig. 69(*b*).)

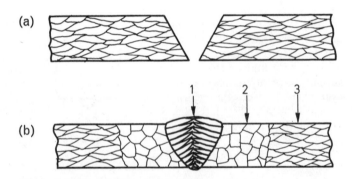

Fig. 69. (a) *Microstructure of cold-rolled aluminium alloy plates* × *1 000.*
(b) *Microstructure of cold-rolled aluminium alloy plates after welding* × *1 000.*

Arrow 1. The weld metal shows an 'as cast' structure. The properties of the weld are therefore comparable to a cast alloy.

Arrow 2. This indicates a region where annealing of the cold-rolled plates has taken place. This lowers the strength and hardness, and cuts down the corrosion-resisting properties.

Arrow 3. During welding the aluminium alloy is not heated sufficiently in this region to cause any changes in the microstructure. It is in the same cold-rolled condition as it was prior to welding.

The non heat-treatable aluminium alloys can have their coarse grained weld metal structure refined by heating to the annealing temperature and hammering. This treatment should not be applied to castings which are never hammered.

Welding processes for non heat-treatable aluminium alloys

Fusion welding of these alloys may be carried out by tungsten inert gas, metal inert gas or oxy-acetylene welding processes.

The metal inert and tungsten inert gas processes produce welds of the finest appearance and quality at very fast speeds. No flux is used, the welds being shielded from the atmosphere by argon gas, and the cathodic cleaning action of the arc ensuring that any oxides formed accidentally during welding are immediately removed. A special claim of these processes is the welding of lap and fillet joints. Because there is no flux used, there is no danger of flux entrapment. Weld designers now have a free choice of use of joints. The fast welding speeds, combined with heat concentration, produce a less heat-affected zone with less distortion when compared with oxy-acetylene welding.

Welds may of course be made with the oxy-acetylene process, with the use of a flux and a neutral flame, but the type of joints made this way are confined to butt and corner. Resistance spot welding is also used on these alloys, but welds are confined to lap joints.

The heat-treatable aluminium alloys

These are aluminium alloys which respond to heat treatment, thereby improving the strength and hardness. Alloying elements are usually added to the aluminium which are capable of going into solution with the aluminium as the temperature is raised. Whilst copper is the principal alloying element added to aluminium, others such as zinc, magnesium and nickel are also used to alter certain properties.

Perhaps the most famous of the heat-treatable aluminium alloys is the duralumin group, composed of aluminium with a maximum of 6% copper. To understand the heat-treatable alloys, it is necessary to consider a basic alloy of 96% aluminium and 4% copper. The microstructure of this material in the 'as cast' condition is

composed of two distinct constituents. (See Fig. 70.) The light
grains contain a solution of 0·3%
copper dissolved in aluminium,
whilst the dark areas within and
around the grains are in the form
of a compound. This compound is
composed of one atom of copper
and two atoms of aluminium

Fig. 70. 'As cast' duralumin × 100.

which are combined chemically,
shown by the chemical symbol $CuAl_2$. This compound is extremely
hard and brittle. If the alloy is now heated to a temperature of
500°C, the $CuAl_2$ compound goes into solution as the temperature
of the alloy rises. At 500°C all of the compound will have been
taken into solution and the entire microstructure is composed of
one solid solution. (See Fig. 71.) If the solution is now allowed to
cool slowly to room temperature the compound precipitates out

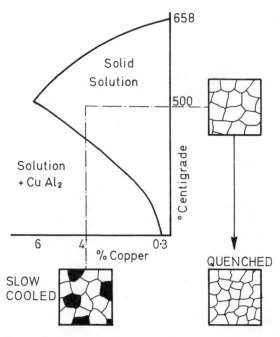

Fig. 71. Heat treatment of duralumin.

126

of solution on reaching the limit of solubility point, as shown in Fig. 71. Precipitation continues until, at 0°C, only 0·3% copper remains in solution. The copper aluminium compound ($CuAl_2$) is again visible in the microstructure. This is a coarse and undesirable type of structure. By reheating to 500°C the copper again goes into the aluminium, forming a solid solution. By quenching in water from this temperature the copper is retained in solution with the aluminium. The solution is unnatural at this temperature and the compound, which should have precipitated out of solution whilst slow cooling, now precipitates out whilst the alloy is at room temperature. The precipitation may take place over a period of six days.

It is regrettable that an ordinary microscope does not show that the copper aluminium compound has precipitated out of solution. The reason for this is that it is distributed in a very finely divided form. The scattering of the compound in this form within the alloy greatly increases the mechanical properties. For example, the hardness of the alloy after quenching is about 60 Brinell. If hardness readings are taken daily, an increase will be noted. After six days, when precipitation is complete, the hardness will be found to be about 120 Brinell. This phenomenon is often referred to as 'age hardening'. The dispersal of the compound throughout the structure has a tightening effect on the atoms which prevents slipping from occurring easily. The effect of this is to increase the tensile strength to about 25 tons per square inch, to increase the hardness and to reduce the ductility. With duralumin in this condition, the alloy is comparable in mechanical properties to mild steel, but it is considerably lighter. The precipitation of the copper aluminium compound takes place over a period of about six days whilst the alloy is at room temperature. The action of precipitation may be speeded up by reheating the alloy which has been previously quenched from 500°C. The temperature of reheating is between 100° and 200°C for a period of only one to three hours. The effect of this is to enable the compound to precipitate out of solution at a much faster rate than when the alloy is left at room temperature. Providing the reheat temperature does not exceed the correct reheating temperature, and that the length of time is correct, then the strength and hardness of the alloy will be increased. An alloy which has been correctly heat treated does not show any visible precipitated compound.

Should the reheat temperature be too high or held at the correct reheat temperature for too long a period, a condition known as 'overaging' or 'annealing' occurs. Evidence of overaging having taken place may be seen in the microstructure if there is visible copper aluminium compound present. (See Fig. 72.) The desirable mechanical properties associated with correctly heat-treated alloys are drastically reduced when 'overaging' takes place.

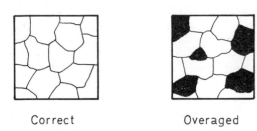

Correct   Overaged

Fig. 72. Effect of reheating on the microstructure × 100.

Welding heat-treatable aluminium alloys

When welding is carried out on duralumin, a most unfavourable heat treatment cycle can result. The strength and hardness of the alloy may be lowered. When fusion welding by the metal inert gas, tungsten inert gas or oxy-acetylene process, heat is applied to the weld area and lost by conduction into the plates being welded. The effect of this on duralumin plates in the heat-treated condition is to cause various changes in the microstructure of the material. Consideration should now be given to the welding of duralumin by the tungsten inert or metal inert gas process. It can be seen from the microstructure in Fig. 73 that there are five different regions of importance to consider when welding is complete.

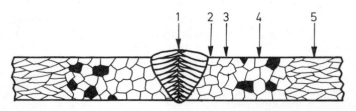

Fig. 73. Changes in microstructure in duralumin after welding by the tungsten inert gas process.

128

Arrow 1. The weld metal deposited shows a coarse 'as cast' structure and consists of filler metal and some base metal fused together. The filler material normally employed contains an amount of silicon—usually 5%. This provides a deposit which, itself, is not heat treatable, but is sufficiently ductile to withstand contraction stresses set up during cooling after welding. If a deposit is not ductile throughout its cooling range, cracking can result when cooling.

Arrow 2. In the fusion zone there is the region where partial melting of the duralumin fuses with the liquid weld filler metal. After solidification of the weld metal, the shrinkage stresses are directly pulling on these fusion faces.

Arrow 3. During welding, the temperature of this region is raised sufficiently to enable the previously precipitated compound to be taken back into solution. The rate of cooling is sufficient to retain the constituents in solution at room temperature. After a period of up to six days, precipitation of the aluminium copper compound takes place.

It should be realized that all the material in this region is in the form of a solid solution immediately after welding. If examined by the microscope immediately after welding, or six days after, when precipitation is complete, there is no apparent change in the microstructure. The reason for this is that the finely divided scattered particles of precipitated compound are too small to be seen, even with the aid of a microscope.

Arrow 4. This region shows a condition of overaging or of a partially annealed state. When the welding heat increases the temperature of this region, some of the compound, which was scattered throughout the grains, is taken back into solid solution. The rate of cooling after this occurs is sufficiently slow to enable visible precipitation of the compound to take place. As stated previously, this condition of overaging is undesirable, lowering the strength and hardness of the material.

Arrow 5. This region is not heated sufficiently during welding to allow any change in the structure to take place, remaining in the original solution of the heat-treated condition.

129

After welding

It is now apparent that regardless of which fusion process is used for welding, five regions will be indicated. If the heat-affected zone base material is required back in its correct original solution heat-treated state, then the entire job must be reheated to 500°C and quenched. This may be followed by precipitation hardening. It must be understood that the weld deposit is hardly affected by the heat treatment, so the weld may not be as strong as the material being welded. Some improvement in the strength of the weld metal can be made by reinforcement or build up.

Welding processes for heat-treatable aluminium alloys

When consideration is given to suitable welding processes, the resistance spot welding process offers the best possibilities. The speed of heating and cooling is so quick that there is little alteration in the microstructure or the properties of the alloy. Therefore the advantage of the process is that solution heat treatment after welding is not required.

Fast welding speeds associated with the metal inert gas and tungsten inert gas processes reduce the heat-affected zone and cut down the risk of fracture from contraction stresses.

The oxy-acetylene welding process should be considered as less suitable than the ones just mentioned. During welding, heat is applied, and conducted over a larger area. This, combined with a slow speed of welding, increases cracking tendencies and contraction stresses.

Magnesiun and its alloys

Far too many welders have grave doubts about the possibilities of welding magnesium and its alloys. It must be understood at once that magnesium and its alloys are readily welded by the metal inert, tungsten inert and oxy-acetylene welding processes. Rumours about fire risks tend to be exaggerated. Except in the thinnest of sheets, any chances of magnesium firing during the course of welding is a direct result of carelessness.

Magnesium is rarely used as a pure metal but is alloyed with a number of additional elements to improve its properties.

Mechanical properties

In the 'as cast' condition, pure magnesium has a low tensile strength of about 8 tons per square inch and is low in ductility, having about 6% elongation value. By alloying, work hardening and correct heat treatment, the tensile strength can be increased to around 20 tons per square inch, whilst the elongation value can reach 15%, thus giving reasonable ductility.

Chemical properties

Magnesium can be alloyed with aluminium, manganese, nickel, silicon, zirconium or zinc. The purpose of adding these elements is to provide an increased resistance to corrosion, to improve the mechanical properties and to improve the weldability. The group of magnesium alloys containing aluminium and zinc are well known by the trade name 'elektron'.

When in the molten condition, magnesium readily forms oxides and nitrides, if allowed to come into contact with the atmosphere. At room temperature, the magnesium alloys are not considered to be sufficiently resistant to corrosion, but this may be improved by applying a chemical surface treatment.

Physical properties

Magnesium alloys are extremely light in weight. They are about two-thirds the weight of aluminium and about a quarter the weight of steel. This factor makes it a favourite material with aircraft

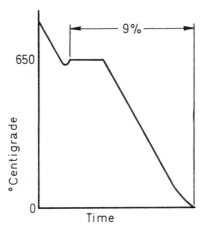

Fig. 74. Volumetric contraction on cooling.

design engineers. Pure magnesium melts at 650°C and boils at 1 110°C. The contraction rate from the commencement of solidification to room temperature is approximately 9% in volume. About 4% of this occurs whilst changing from a liquid state to a solid state. As this occurs, heat is lost without any decrease in the temperature, i.e., in the latent heat period. (See Fig. 74.)

The high conductivity of magnesium and the expansion rate make the preheating of thick sections before welding essential. The low specific heat and latent heat values and melting point mean that only a small quantity of heat is required to raise magnesium to welding temperature.

Problems during and after welding

The welding problems encountered with magnesium alloys are quite similar to those of aluminium, in that they have the following characteristics in common: oxidation, high thermal conductivity, high thermal expansion, low melting point, cracking and, if used, flux removal.

Oxidation

Before welding, magnesium alloys should be given a similar cleaning treatment to that required when welding aluminium. Oxidation can be prevented by the use of a flux when oxy-acetylene welding, or the protection of argon gas combined with the cathodic cleaning action when using the tungsten inert gas or metal inert gas processes.

Thermal conductivity: Thermal expansion

During welding, heat is lost from the weld area, this results in a considerable amount of uneven expansion throughout the alloy. Preheating assists in balancing the expansion and contraction stresses. The contraction rate is very high and, if allowance is not made, severe distortion may result or even cracking.

Low melting point

Very often the addition of alloying elements lowers the melting point of magnesium alloys. Because a preheat may have been used, coupled with a low melting temperature, welding speeds may have to be fast.

Cracking

Magnesium is more liable to crack than aluminium. Not only may cracking take place during welding through stresses being set up, but also stress cracks may occur after a considerable lapse of time. To prevent this from occurring, stress relieving after welding is essential. Stress relieving can be carried out by heating the component to a temperature of 260°C for up to an hour. If the stress relieving temperature is too high or if the period of time is too long, serious detrimental effects result, such as grain growth, or a condition of overaging in the case of solution heat-treatable magnesium alloys.

After welding, straightening of certain sections may be needed. If this is carried out whilst the alloy is at room temperature, cracking results. When straightening the magnesium alloys, it is necessary to reheat to a temperature of between 200° and 300°C.

Flux removal

When oxy-acetylene welding magnesium alloys, a positive guarantee of flux removal is necessary. Similar to aluminium, the process is somewhat limited to the use of butt joints and corner joints. All flux residue should be removed after welding by brushing in hot running water. Then the component should be immersed in a solution of 5% dichromate and hot water.

Selection of welding process for magnesium alloys

Magnesium alloys are not welded by the manual metal-arc process. The preferred processes of welding magnesium alloys are the metal inert and tungsten inert gas processes. These are capable of protecting welds from the undesirable atmospheric gases which readily combine with magnesium. The welds made by these processes are of a high quality and appearance.

The oxy-acetylene process provides satisfactory results when used, but, as previously stated, it is limited to butt and corner joints, as well as needing the complete removal of flux residue after welding.

The resistance spot welding process finds many applications on lap joints required on relatively thin sections, especially those used in the aircraft industry.

Unfamiliar terms used in this chapter

Ore: A mineral from which metal may be extracted.

Electrical reduction of oxide: Method of rendering certain metals pure.

Precipitation hardening: Increase in hardness obtained by inducing a dissolved compound to come out of solution in a finely dispersed form.

Density: Mass of matter per unit volume.

Pickling: Chemical removal of an oxide.

Cathodic cleaning: Removal of oxide by electrical means.

Consumable electrode: An electrode which is continually fed into a weld pool and becomes part of the solidified weld deposit.

Limit of solubility: Dissolved to a maximum amount for a given temperature.

Slipping: The ease with which a row of atoms may move over each other. A ductile material will slip a great deal without fracture.

Chapter review questions

What is meant by cold working?

State the annealing temperature of aluminium.

Name two ways in which the mechanical properties of aluminium may be improved.

How can flux residue be removed when aluminium alloys have been welded by the oxy-acetylene welding process?

What would be the effect of welding upon the microstructure of duralumin?

Why are oxy-acetylene welds not recommended for fillet and lap joints when welding aluminium and magnesium and their alloys?

What is meant by overaging?

Why is a 5% silicon filler rod recommended when welding?

How many days would be required to ensure that precipitation was complete when age hardening?

How may the properties of magnesium be improved?

Which is considered to be the preferred process when welding the aluminium and magnesium alloys?

What important physical properties should be considered when welding magnesium alloys?

Why do magnesium alloys find several applications in the aircraft industry?

CHAPTER 12

Copper and its Principal Alloys

Copper

Copper is a metal which is widely used in industry nowadays. When it is in the pure form, it is used in electrical wiring, piping, chemical plant and has many other applications. Years ago, some welders were doubtful about the possibilities of joining copper, possibly through previous failure. It must be stated immediately that a lot of copper is welded nowadays and the amount is likely to increase. Metal inert gas (M.I.G.), tungsten inert gas (T.I.G.) and oxy-acetylene have been found to be suitable for joining certain fusion weldable grades. Resistance welding is used in joining very thin copper sheets and copper wire. All grades of copper, whether suitable for fusion welding or not, are capable of being bronze-welded. Successful welding of copper requires an appreciation of the physical, chemical and mechanical properties of the metal. The principal factors affecting weldability are as follows:

Thermal conductivity

The thermal conductivity of copper is high, roughly about five times that of steel. This means that heat is lost rapidly from the weld area. It is necessary when welding to supply more heat to make up for this loss; the result of this is to cause expansion of large volumes of metal followed by subsequent contraction on cooling. This may cause severe shrinkage stresses. Methods which may be adopted to reduce or compensate for heat losses from the weld zone are:

    (1) Covering or backing the work as far as possible with an insulating material, such as asbestos sheet or firebrick.

(2) Preheating both the fusion zone itself and considerable adjoining areas. The oxy-acetylene blowpipe is a very handy method of doing this, but on large work, coal gas burners and electrical type heaters may be found to give more uniform heating.

(3) A more intense source of heat for welding can be used, for example, the M.I.G. or T.I.G. processes. The advantage here is that the melting of the metal is quicker and lower preheat temperatures are possible. In oxy-acetylene welding the quantity of heat can be increased by the use of larger size welding nozzles. When resistance welding, an increase in current may offset heat losses.

Thermal expansion

Copper has a coefficient of expansion about 50% higher than mild steel. When fusion welding, serious consideration should be given to its effects. The heating of the work by the arc or flame is usually applied locally to one place; it is an uneven heat and therefore leads to distortion. The severe amount of contraction on cooling, particularly with large articles, may lead to serious shrinkage stresses. The majority of metals suffer a reduction in tensile strength as the temperature is raised and copper is one of these metals. The tensile strength of copper falls considerably as the temperature is raised above 250°C. If the copper being welded, or the weld metal, has low strength and ductility over any range of temperature below the solidification point, the contraction stresses may exceed the tensile strength. The result of this will be cracking, known as hot shortness.

Precautions which can be taken against such trouble when welding may be as follows:

(1) Confine serious heating to the immediate neighbourhood of the fusion zone by welding rapidly, using a sufficiently intense source of heat.

(2) Preheat the entire component to a even temperature of between 200°C and 600°C dependent upon the welding process to be used. After welding, the component should be allowed to cool evenly and gradually.

(3) Before welding, any device which attaches the pieces to be welded to any rigid structure should be removed. These may be in the form of bolts or rivets. The reason for this is that they restrict the contraction of the component whilst it is cooling.

(4) After welding, the weld zone should be hammered whilst it is at dull red heat or above. The advantages of this are that the grain structure is refined and there is an increase in the tensile strength and impact value.

Electrical conductivity

Copper is an excellent conductor of electricity. Because of its high conductivity, it allows current to pass through without an appreciable increase in heat. This factor restricts the resistance welding of copper to the joining of relatively thin sections. Even with thin sections, high current densities are required to raise the parts to be joined to the temperature required for welding. The best results when resistance welding copper are obtained when using the upset butt and flash butt methods. The ends to be joined are heated locally by the passage of current which is confined to the cross section of the two ends, these being in contact. Because the heat created is between two small cross-sectional ends, the area of the conduction path is also small, which in turn reduces heat losses by conduction when compared with other resistance welding methods.

Mechanical properties

Copper after annealing has a tensile strength of about 15 tons per square inch and an elongation of about 30%, indicating good ductility. If cold working is carried out, the tensile strength can be increased to over 30 tons per square inch, but its elongation is reduced according to the amount of cold work. The effects of cold working are removed when welding is carried out on copper on all the parts that have been heated above recrystallization temperature.

Chemical properties

Copper is a metallic element with a melting point of 1 083°C. It may be regarded as one of the purest metals that it is possible to obtain commercially, exceeding 99·90%. There are two main groups of copper, namely tough pitch copper and deoxidized copper.

Tough pitch copper

Copper which contains a small amount of oxygen is known as tough pitch copper. The percentage of oxygen is only 0·05% and is found in the form of cuprous oxide. When a piece of tough pitch copper in the 'as cast' state is viewed under a microscope, it will be seen that the grain boundaries show a netlike pattern of cuprous oxide. This makes the material brittle and weak. (See Fig. 75.)

Fig. 75. 'As cast' tough pitch copper × 100.
Fig. 76. Hot and cold-worked annealed tough pitch copper × 100.

In the course of hot working followed by cold working and annealing, the structure is recrystallized and the cuprous oxide is now found distributed evenly throughout the metal. It can be seen under the microscope as tiny particles in the grains. The effect of the cuprous oxide in the grains is to increase the tensile strength and ductility of the metal. (See Fig. 76.)

If, for any reason, tough pitch copper is melted, on solidifying, the structure reverts to the original 'as cast' condition with its inherent weakness. The beneficial effects of the cuprous oxide, having been evenly distributed throughout the metal, are lost. This is one of the reasons why tough pitch copper is not fusion welded.

Steam reaction or gassing

If tough pitch copper is heated to a temperature of 915°C, the cuprous oxide forces its way to the grain boundaries. If there are any reducing gases such as hydrogen or carbon monoxide present, reduction of the cuprous oxide takes place. Consider the effect of hydrogen; this readily combines with the oxygen in the cuprous oxide and the product of this combination is pure copper and steam. On being formed, the steam is rapidly expanded by the source of heat and, being at the grain boundaries, will force them apart. This is shown by the formation of cracks in the metal. Carbon monoxide also reduces the cuprous oxide, forming carbon dioxide and releasing pure copper. Both these two reducing gases can be found in the reducing zone of any oxy-acetylene flame; therefore, when welding, the difficulties associated with gassing cannot be avoided.

138

Bronzewelding is the recommended method of joining tough pitch copper. The reason for this is that the temperature required for bronzewelding is between 800° and 900°C. At this temperature, the distribution of the cuprous oxide is unaffected.

Deoxidized copper

This is produced by the addition of between 0·04% and 0·15% phosphorus; silicon may also be present up to 0·15%. The benefit gained from the addition of these elements is that the copper is entirely free from oxygen. Deoxidized copper is readily weldable by fusion methods such as the M.I.G., T.I.G. and oxy-acetylene welding processes. The reason for this is that there is no cuprous oxide in the copper and therefore there are no gassing difficulties during welding. Welding of deoxidized copper can also be carried out on thin sections by the resistance methods of welding.

Effects of fusion welding on cold-rolled deoxidized copper

When two pieces of cold-rolled deoxidized copper plates are welded, it is essential to understand what may take place in the internal structure of the metal when heat is added.

Perhaps an explanation should begin by considering the macro-structure of a vertically cast deoxidized copper bar. It will be noticed in Fig. 77(*a*) that the structure is coarse and columnar. This structure is weaker and less ductile than would be the case of copper in the wrought or annealed condition. But this does not mean that it is not ductile enough to withstand cold work, in fact this is the next operation.

(a)   (b)

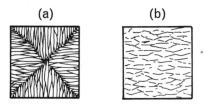

Fig. 77. (a) *Macrostructure of a cast copper bar* × *4.*
     (b) *Microstructure of heavily cold-worked copper* × *100.*

The cold working can be carried out by hammering, rolling or drawing, making the copper brittle and harder. The general direction of the cold working can be seen in Fig. 77(*b*). The grain boundaries and the grains are broken up, resulting in general distortion.

When copper has been work hardened, it can be returned to its soft and ductile state by annealing. This may be carried out by heating the copper to a sufficient temperature for recrystallization to occur. During annealing, new crystals form from the remnants of the distortion, at a temperature between 250° and 600°C. After heating to the annealing temperature, the copper is then allowed to cool slowly, although it may be quenched. It is sometimes thought that the quench is essential, but this is quite wrong. Regardless of the rate of cooling, recrystallization takes place before the copper begins to cool.

Figure 78 shows pure deoxidized copper which has been cold worked and annealed. The newly formed crystals are equiaxed and the metal is soft and ductile. It may be seen that these newly formed crystals bear no resemblance to the coarse columnar crystals of the 'as cast' structure. Further cold working can be carried out on

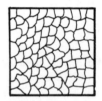

Fig. 78. Cold-worked and annealed deoxidized copper.

copper if desired, and by subsequent annealing it can be restored again to a fine grained equiaxed structure.

By reconsidering the coarse 'as cast' structure in Fig. 77(*a*), it should be realized that the only way in which this may again be produced is by melting and subsequent solidifying. Melting occurs when fusion welding is carried out. On solidification, the resulting weld metal shows an 'as cast' coarse structure. A macrostructure of a weld made between two pieces of cold-rolled deoxidized copper can now be considered. Using the tungsten inert gas welding process, a butt weld is made between the plates; the macrostructure of this is shown in Fig. 79.

Fig. 79. Macrostructure of T.I.G. weld in deoxidized copper × 4.

The macrostructure shows three distinct zones, the deposited weld metal, the annealed zone and the cold-rolled copper unaffected by heat. The weld metal is molten during deposition, and on solidification it has a coarse columnar 'as cast' structure. Heat is lost from the weld by conduction into the plates being welded. The areas near the weld are heated to above the recrystallization temperature and annealing of the cold worked plates takes place. Further away from the weld is the metal which is never heated to the recrystallization temperature; it therefore remains in the cold-worked condition. A weld which has been made by this process is acceptable in the welded state for many purposes. However, if a fine equiaxed grain structure is required, reheating and hot working have the desired effect.

Successful welding depends upon the welder having a thorough understanding of the properties of copper and the effect that welding heat has upon these properties. A code of procedure must be adopted and rigidly adhered to before, during and after welding.

Brasses

The most common alloying element added to copper is zinc. It is found in quantities of up to about 40% among the commercial brasses in use today. These brasses can be conveniently grouped into two classes, each possessing distinctly different properties and having their own particular applications. They are:

(*a*) The brasses that contain up to 30% zinc, remainder copper.

(*b*) The brasses that contain 40% zinc, remainder copper.

The microstructure of a brass containing up to 30% zinc shows, in the annealed condition, a solid solution of zinc entirely dissolved in copper. (See Fig. 80.) Two well-known types of brass are composed of an entirely solid solution, these being *gilding metal* and *cartridge brass*. Gilding metal contains between 5% and 20% zinc. It is usual to write down the composition when specifying brass, e.g. 80/20, 85/15, 90/10, 95/5, the copper content being shown first. Cartridge brass is usually written as 70/30. Referring to this class of brasses only, with

Fig. 80. Brass containing 30% zinc or less.

141

increases in the zinc content, the tensile strength, hardness and ductility increase. The cartridge brasses are very suitable for articles which may be subject to cold work during manufacture.

If the microstructure of a 60/40 brass is considered, it is found that there are two constituents to be seen, which indicates that another solution has been formed. This is, in fact, a zinc-rich solution; its characteristics are hardness and brittleness. How- ever, when the zinc-rich con- stituent is present in brass it gives better strength and hardness. The 60/40 type brass is known as Muntz metal or yellow brass. (See Fig. 81.) This type of brass should be hot worked, although a small amount of cold working may be carried out.

Fig. 81. Yellow brass, 60/40 composi- tion. Note : it is a mixture of two distinct solutions.

Brasses of the 60/40 composition which contain small percentages of additional alloying elements, such as manganese, tin, chromium or aluminium, are generally known as *high tensile brasses*. Some- times about 2% to 4% lead is added to brass to improve its machinability. The welding problems are created by the amount of zinc present and also the lead content, if any. The effect of the other alloying elements on welding does not cause serious welding problems, though an increase in the brittleness of the alloy may result. (Except in the case of aluminium: see the section on the welding of aluminium bronze.)

Mechanical properties

The tensile strength of annealed 70/30 brass is around 22 tons per square inch, whilst its ductility is very high and its elongation is 70%. The average figures for the 60/40 brasses are around 25 tons per square inch for tensile strength and an elongation of 38%.

Chemical properties

Copper melts at 1 083°C on its own, whilst zinc melts at 420°C and boils at 906°C. To carry out welds successfully, zinc losses must be kept to a minimum. In vapour form zinc readily combines with oxygen from the atmosphere forming white clouds of zinc oxide. One effect of zinc leaving the weld pool in the form of vapour is the creation of gas being trapped during the solidification of the weld metal.

Lead present in brass is detrimental to welding. It does not go into solution with brass and therefore leads to cracking, whilst hot, and porosity. Brasses containing lead should not be welded, but joined by a brazing method.

Physical properties
The thermal conductivity of brass is much less than that of copper, but it is still a good conductor and thicker sections must be preheated. The expansion rate of brass is very high, which can lead to distortion. It is also possible to get a certain amount of distortion if welding is carried out on brass which has been previously cold worked, stress relieving occurring due to welding heat.

Welding difficulties

Welding is carried out at a temperature range of about 850° to 1 000°C. It is therefore necessary to have a definite control over the zinc activities and losses. The weld pool should not be overheated; the higher its temperature, the greater the zinc losses. Some measure of success in preventing zinc losses may be obtained by adding silicon to the filler wire. The purpose of this is to form an oxide film on the surface of the weld pool, thus preventing zinc losses or gases being absorbed by the weld. Quite a large percentage of brass is purchased in the cold-worked condition. Therefore the brass will contain internal stresses and it is advisable to remove the effects of cold working before welding. This may be done by evenly preheating to 300°C which relieves the stresses. Owing to residual stresses being caused by contraction when cooling after welding, a postheat should be given to the component, the temperature being 300°C. If welding is carried out on brasses containing more than 20% zinc, failure may result if stress relieving by preheat is not carried out. This also applies equally if postheating does not take place. The term referring to cracking due to stresses in brasses is known as 'season cracking'.

Welding processes for brasses

The oxy-acetylene welding process is preferable when welding brass. The use of an oxidizing flame and the introduction of silicon over the weld pool from the filler wire forms a film of silicon oxide on the surface, reducing zinc losses. The filler wire is composed of copper and zinc, along with a small percentage of silicon, and therefore produces a homogenous weld.

When using the manual metal-arc process on welding brasses, an electrode should be composed of aluminium bronze or phosphor bronze. Zinc will be lost by volatilization if any attempt is made to transfer it across from an electrode to the work in arc welding. The manual metal-arc produces satisfactory welds, but the deposit is of a different composition to the parent material.

The T.I.G. and M.I.G. welding processes are suitable for the welding of brasses, the filler wire preferably being of copper silicon or copper tin alloy. Zinc losses from the molten parent material during welding are controlled, to a satisfactory extent, by the use of low current values.

Phosphor bronze

Alloys of copper and tin are termed as tin bronze, and if a small percentage of phosphorus is added, they are known as phosphor bronzes.

Mechanical properties

The average tensile strength of phosphor bronzes is about 15 tons per square inch, the elongation percentage being about 25%. In the cold-worked condition, they possess the desirable properties of being fatigue resistant, hard and tough.

Chemical properties

The chemical composition is between $1\frac{1}{2}\%$ and 10% tin, with a phosphorus content of 0·1% to 0·3%, the base metal being copper. The purpose of the phosphorus content is to act as a deoxidizer. An important feature is the corrosion resistance of these alloys, particularly in acids and sea water.

Physical properties

The tin melts at the low temperature of 232°C and boils at about 2 300°C. The bronzes have a wide range of temperature of solidification. For example, a 10% tin alloy starts to solidify on cooling at over 1 000°C, but solidification is not complete until about 850°C is reached. This long temperature range of solidification can

144

lead to shrinkage cracks, so, when welding, it must be cooled through this range as quickly as possible. Another reason for rapid cooling is the fact that phosphor bronze, like other copper alloys, is subject to hot shortness.

Welding difficulties

Welding should be carried out in the shortest possible time, because contraction stresses whilst at elevated temperatures may cause cracking. A low preheat of about 250° to 300°C increases the speed of welding. Copper backing bars conduct heat away from the weld area which assists in rapid solidification of the weld metal. The weld metal deposit has a coarse grain structure. This can be improved by peening whilst hot, followed by annealing at about 500°C.

Welding processes for phosphor bronze

The preferred method of welding phosphor bronze is the manual metal-arc process. The reason for this is that high welding speeds, fast cooling rates and localized heat keep the shrinkage stresses to a minimum. A special claim for this process is that few gases are entrapped in the solid weld. An electrode should contain up to 0·5% phosphorus to ensure deoxidation.

All resistance welding processes are suitable for the joining of phosphor bronzes. The reason for this is that the conductivity of the material is quite low. An advantage of the resistance welding processes is that only a small area around the weld is affected by the heat of welding.

The oxy-acetylene welding process is to be regarded as the least used method of welding the phosphor bronze alloy. The reason for this is the high heat input, which spreads over a wide area and results in an excess amount of contraction stresses which may produce cracking. The slow rate of cooling of an oxy-acetylene weld may result in gases being absorbed, thus causing porosity.

Aluminium bronze

There are two types of aluminium bronze available, these are:

Non heat treatable

This may contain between 4% and 7% aluminium dissolved in copper, forming a solid solution. It is an alloy which may be hardened by cold working. Care should be taken not to carry this out too much because it work hardens rapidly and cracking results. The effects of cold working may be removed by annealing at a temperature of between 400° and 700°C.

Heat treatable

This contains an average of 10% aluminium. Some of the aluminium is dissolved in solution with copper, whilst some is found as an additional hard brittle constituent when slow cooled. This type of aluminium bronze is heat treatable in a similar manner to steel. (See Fig. 82.) At about 900°C the alloy consists of a solid

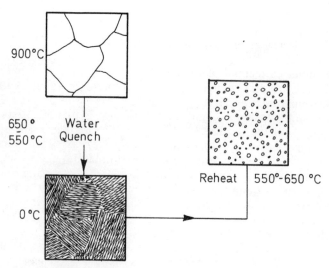

Fig. 82. Heat treatment of 10% aluminium bronze.

solution, similar to austenite in steel. When quenched rapidly, a needle-like structure appears similar to martensite not only in appearance but in its characteristics, being hard, brittle and of low ductility. For improved toughness, it is reheated or tempered at about 550° to 650°C. There will be a slight reduction in hardness and strength, but a great increase in the ductility. Typical figures for this alloy in the heated, quenched and tempered condition are as follows:

U.T.S. (ton/in²)	Elongation %	Brinell hardness no.
39	25	176

Welding difficulties

Non heat-treatable alloys are prone to cracking whilst hot. The alloy suffers from severe grain growth in the heat-affected zone during welding and, because of this, welding should be completed with a maximum of two layers when arc welding. On cooling between 700° and 500°C the alloys are prone to hot shortness and should be quickly cooled through this range.

The heat-treatable type of aluminium bronze is a better welding proposition. The reason for this is that the hot shortness range is much smaller, multi-layer welding can be carried out, and the cracking tendency is reduced. Welding destroys the properties produced by previous heat treatment so that reheat treatment is necessary after welding.

Welding processes for aluminium bronzes

The oxy-acetylene welding process is not recommended for the welding of these alloys. A limited amount of success has been achieved, but difficulties in effective fluxing to remove aluminium oxide is a constant source of trouble. If this process is used, filler wire should be of the same composition as the parent material.

The manual metal-arc welding process offers better results, electrodes of matching composition being used. The deposits have reasonable strength and ductility at elevated temperatures and they are able to withstand the contraction stresses.

The metal inert and tungsten inert gas welding processes produce consistently high quality welds and must therefore be considered the preferred processes. There are no fluxing difficulties, the argon providing a shield from the atmosphere, and any small amounts of oxidation occurring are removed by the cathodic cleaning action of the arc.

Welding Science and Metallurgy

Unfamiliar terms used in this chapter

Cuprous oxide: Copper oxide.
Drawing: To stretch out lengthways.
Machinability: Capable of being shaped, drilled, turned or ground.

Chapter review questions

What are the practical uses of copper?
Name the most suitable welding processes for joining copper.
Why is fusion welding of tough pitch copper not recommended?
How does deoxidized copper differ from tough pitch copper?
Why is copper preheated when welding by the oxy-acetylene process?
How may copper be hardened?
What is meant by steam reaction?
How is annealing carried out on copper?
What is the composition of cartridge brass?
When welding certain brasses, which elements are detrimental to good welding?
Why is silicon added to the filler rod when welding brass by the oxy-acetylene process?
Why is zinc excluded from covered electrodes?
Why is phosphorus used in tin bronzes?
When welding phosphor bronze, why should welding be completed as quickly as possible?
Compare the similarities of heat-treatable aluminium bronze and a heat-treatable carbon steel.
What are the recommended welding processes for aluminium bronzes?

CHAPTER 13

Cutting of Metals and Alloys

The rapid combustion of iron in oxygen takes place when the iron is heated to a high temperature. Lavoisier noted this fact more than one hundred and fifty years ago. He also suggested that the oxide formed was more fusable than the iron and, being detached, could be removed as the combustion proceeded.

Combustion of iron in oxygen

Most metals oxidize when they are in contact with the oxygen present in the atmosphere. This is a slow combustion which continues until the layer of oxide on the surface is dense enough to protect the rest of the metal from the action of the atmospheric oxygen. If a piece of polished iron or steel is placed in a position where it is exposed to the atmosphere, a thin layer of rust (or oxide) forms on the surface, therefore slow combustion has taken place.

On the other hand, it is known that oxidation is much more intense and rapid as the temperature of the metal is increased. The problem of oxidation has constantly to be overcome when fusion welding is being carried out.

If a thin piece of iron or steel is considered, such as a $\frac{1}{8}$ in diameter welding rod, it is possible, by first raising the rod at one end to red hot, to bring about rapid combustion. This may be done by plunging the heated end into a jar containing pure oxygen, the iron rapidly burns in contact with the oxygen. The oxide of iron which is formed is detached from the metal, and projected on all sides in a molten state which rapidly solidifies. (See Fig. 83.)

What is the 'mechanism' of this rapid combustion of iron in

149

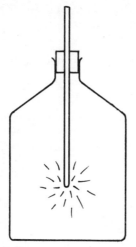

Fig. 83. Rapid combustion.

oxygen? The oxidation commences at the part which has been previously heated to redness; because at this temperature the reaction takes place readily. The combustion of this portion of iron disengages heat, a portion of which is absorbed by the neighbouring part. This is sufficient to raise it to red heat, so that it in turn burns and the reaction is progressively propagated throughout the metal. The oxide, when formed, has a much lower melting point than that of the metal, and is detached leaving the iron at the surface continually bare.

Copper, brass, bronzes and aluminium do not burn in oxygen in the manner which has been described above. This is because, when in contact with oxygen, their oxides have a melting point equal to or higher than that of the metal, and are prevented from being detached.

Therefore iron and low carbon steels are alone amongst the ordinary materials which oxidize in a continuous manner. This is because the oxide of iron produced is eliminated as it is formed, in the molten state. It is useless to attempt to cut by oxygen alone any other metal or alloy which does not possess this property.

Oxy-fuel gas cutting

The introduction of oxy-fuel gas cutting in the early 1900s was an important step forward in industry, and it is still one of the most widely used methods of severing metals. Mild steel plates may be cut from the thinnest available sections to any thickness required. Because of the simplicity and ease of operation of the process, complex shapes are readily cut in a time which compares well economically with other methods of cutting.

In order to produce a successful cut by the oxy-fuel gas process, a small area on the top surface of the mild steel plate is heated to red heat. At this temperature, as was seen in Fig. 83, iron rapidly oxidizes. The preheat of the plates to ignition temperature (red heat) is obtained by the combustion of a fuel gas, such as acetylene with oxygen. At this temperature, a stream of high velocity pure oxygen is projected on to the surface where heated; this is the cutting stream of oxygen and is independent of the oxygen in the preheat flame. Combustion rapidly takes place forming molten iron oxide, which is driven away as it is produced by the energy of the high pressure stream of oxygen. This action continues because the combustion extends throughout the thickness of the material to be cut. (See Fig. 84.)

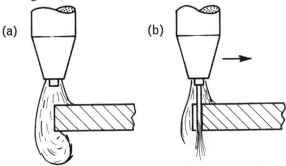

Fig. 84. (a) *Preheat before cutting.* (b) *Oxygen directed on to heated surface.*

In the experiment where the $\frac{1}{8}$ in diameter mild steel welding rod is plunged into the jar of oxygen, the heat of the reaction is sufficient to maintain the temperature necessary for the oxidation of the adjoining portion, and so on. In the process of cutting it is slightly different. In reality, the conductivity of the metal to be cut absorbs a considerable part of the heat given to the metal, and the temperature is not maintained in the part to be cut to the degree necessary for oxidation. For the reaction to proceed it is therefore necessary to add constantly the amount of heat necessary to maintain the part of the metal to be cut at red heat. This heat is only necessary on the top surface layer of the metal and is supplied from the preheat flame.

Heavy sections of mild steel are cut with comparative ease, for, even in this case, it is only the top surface layer which needs to be maintained at the ignition temperature by the preheat flame. When the iron becomes oxidized, heat is given out due to the chemical reaction which takes place. When the liquid oxide is forced through the thickness of the plate, the heat from the reaction raises the metal beneath it to ignition temperature where it combines with

151

and is removed by the oxygen stream continuing the cut; this is progressively propagated throughout the thickness of the material.

In theory, 4·6 cubic feet of oxygen gas will combine with one pound of iron to form magnetic iron oxide (Fe_3O_4). However, in practice, it has been found that the oxides obtained from cuts in iron and steel with a jet of oxygen may not necessarily be of magnetic iron oxide, but of other oxide compositions. However, these other compositions are still removed by the high velocity oxygen stream.

Fuel gases used for cutting

There are several fuel gases which may be used in conjunction with oxygen for the preheating of the material to be cut, including acetylene, propane, coal gas and hydrogen.

Acetylene

Acetylene gas is a chemical compound of carbon and hydrogen. It is probably the most widely used of all the fuel gases, for when acetylene is mixed with oxygen and ignited, the flame produced has a temperature higher than that of any other fuel gas burning with oxygen. The advantage of this high temperature flame is that thick or thin sections of mild steel can be cut at very fast and economical rates. It is also easy to ascertain that the flame is neutral.

Propane

This can be supplied in the form of a liquid in equilibrium with gas under pressure in a cylinder. As with acetylene, propane gas is a hydrocarbon compound but in this case it requires more oxygen to ensure complete combustion than is required for complete combustion of acetylene.

Coal gas

This gas is usually supplied via a non-return valve from the town mains supply, or it may be obtained compressed in cylinders, but this is more expensive. Coal gas is not usually used on site work because its uses are confined to clean plates and machine cutting in a workshop. The cut is very easily 'lost' through the low temperature of the preheat flame, which is only 2 000°C.

Hydrogen

Hydrogen is capable of mixing with oxygen to provide a preheat flame, but the cost of this gas is quite high and its use is therefore restricted. The products of combustion are non-toxic; this is because there is no carbon available to combine with oxygen forming carbon monoxide or carbon dioxide. When cutting has to be carried out in confined spaces, the use of hydrogen should therefore be considered.

Hydrogen was once used exclusively for underwater cutting operations as it can be compressed safely to high pressures and then regulated to overcome the pressure exerted by the water. Nowadays other gases are replacing hydrogen for underwater cutting, for example, methane.

Oxygen purity

When oxygen is used for cutting purposes, the best results are obtained only when the gas is of high purity—not less than 99·5% pure. Oxygen is obtained by liquefying the atmosphere. Other gases, such as nitrogen or argon, may be present as impurities and if they exceed 0·5% the cutting speed will be considerably reduced and a greater volume of cutting oxygen will be consumed.

The effect of cutting on steel

During the course of oxy-fuel gas cutting, the largest quantity of heat is produced by the iron oxygen reaction. The heat produced is far in excess of the heat required to continue the cut to the full depth of the material. Some of the heat is conducted into the material being cut, the area adjacent to the cut edge is therefore heated through the transformation range of the material. As the cut is continued, the colder adjoining material causes rapid cooling of the steel above the critical range. The rate of cooling from the high temperature determines the arrangement of the constituents of the steel and, correspondingly, its physical properties. It is true to say that when steel has been cut by the oxy-fuel gas process there is always a heat-affected zone at each of the cut edges. This may be described as follows:

  (1) Regardless of the carbon content of the steel, there is evidence of carbon migration to the edge of the cut. This carbon may be from the steel being cut or from carbon being released during the oxidation reaction.
  (2) The rapid cooling rate exerts a quench effect on the heated area resulting in hardness along the cut edge. In the case of mild steel the cut edge is slightly harder and appreciably

L

tougher than when fully annealed. This hardness decreases progressively away from the cut edge.

When higher carbon or alloy steels are considered, the depth of hardness along the cut edge may be as much as $\frac{1}{16}$ in.

Preheating

Generally, materials cut by the oxy-fuel gas processes are preheated to the ignition temperature by the preheating flame of the cutting blowpipe. This is a local preheat and is applied for the purpose of keeping the top edge of the plates being cut at the ignition temperature. There should be no confusion between this local preheat and a general preheat which is necessary when cutting certain alloy or high carbon steels. When cutting steels containing 0·3% carbon or less, a general preheat is unnecessary. With further increase in the carbon content and the addition of alloying elements, difficulties may arise as a result of heat created during cutting. Heat can be lost rapidly by conduction into the material being cut and the cooling rate of the cut edge will be fast. There is a possibility of martensite being formed along the cut edge, resulting in excessive hardness and, perhaps, cracking, through the effect of contraction stresses. To overcome these difficulties, general pre-heating of the material should be carried out. The chances of cracking along the cut edge are further increased when cutting heavy sections or when intricate shapes have to be cut. The temperature of the preheat is quite low, between 100° and 300°C. If the material to be cut is held between this temperature range, the quench effect is reduced and the cooling rate is therefore decreased. There will be no martensite formed along the cut edge, and the hardness and brittleness are reduced, this in turn cutting down the cracking tendencies.

Advantages of preheating

These are as follows:

(1) The preheating of the material by the cutting blowpipe to the ignition temperature is completed in a shorter time and also the actual speed of cutting is increased.

(2) The cooling rate is reduced, preventing excessive hardness along the cut edge.

(3) The contraction on cooling is evenly balanced throughout the material so that the distortion is reduced.

154

The effect of alloying elements when cutting steels

Basically, steel is composed of iron and carbon. Both of these elements readily oxidize and have already been considered. However, the addition of other alloying elements increases the difficulties of cutting by the oxy-acetylene process. The following alloying elements should be considered:

Manganese
Steels containing up to 1·6% are readily cut, and the cut edge will be quite good. Steels containing up to 14% manganese can be cut but should be rapidly cooled to prevent the formation of manganese carbide, with its embrittling effects, along the cut edge.

Nickel
Up to 3% nickel in steel increases the hardness of the material, but these low nickel alloy steels are easily cut by the oxy-fuel gas process. Even up to 7% nickel in steels still permits a reasonable cut to be made. The presence of 8% nickel or more renders the material to be austenitic at room temperature. Due to nickel not readily oxidizing when at or above this percentage, successful cuts can only be made by using processes that do not rely on an oxy-fuel gas flame alone for preheating to cutting ignition temperatures.

Silicon
Steels containing up to 4% silicon are cut quite readily by the oxy-fuel gas process. If a steel alloyed with silicon and containing a high percentage of carbon and manganese is required to be cut by this process, then preheating and subsequent postheating should be carefully carried out, the post heat temperature being 300°C.

Chromium
If chromium is present in steels up to 5%, it does little to interfere with the cutting of the material. Above 5% difficulties are encountered; this is because at high temperatures chromium combines readily with oxygen to form refractory oxides which may be considered unfusable. Up to 10% chromium in steels can be cut, provided use is made of preheating.

Heat treatment after cutting
Even though careful consideration has been given to preheating and alloying elements, metallurgical changes can take place in the materials which have been cut. Subsequent annealing, normalizing or stress relieving may be required to remove any undesirable structures which are present.

Powder injection cutting

The information regarding cutting by oxygen and a fuel gas *alone* indicates that the process is somewhat limited to the cutting of materials which oxidize at a temperature that is lower than the melting point of the material. Thus the process is confined to the cutting of ferrous materials which contain a limited amount of alloying elements. In fact, until about 1944, materials such as stainless steel, aluminium and copper, etc., were somewhat difficult to cut and shape by the available processes.

Moderate results were obtained on stainless steels by the application of certain rough methods. One of these was to place a piece of mild steel on the top surface of the plate to be cut. The oxy-acetylene cutting blowpipe was used and the cut was carried out through both plates at the same time. A similar effect to this was produced by depositing two or three mild steel weld beads by the manual metal-arc process over the line of the cut. The oxy-acetylene preheat flame was directed on to the weld deposit and, on reaching ignition temperature, a jet of high velocity cutting oxygen was directed on to this point and a cut was made. Certain successes were gained by holding any oxy-acetylene preheat flame over the area to be cut. A mild steel welding rod was held in the flame and, when at ignition temperature, the cutting oxygen jet was brought into contact with it. Excessive heat was given out by the rapid oxidation of the iron, thus a rough cut was made. These methods of cutting were not of a standard that could be relied upon and a considerable amount of time was wasted when cleaning the cut edge.

These rough methods, however, were quite definitely the fore-runner of the modern powder injection process. By the use of this cutting process, the otherwise difficult to cut materials can be cut, bevelled and shaped with successful results.

Powder injection principle

In this process an iron-rich granular powder is conveyed into the oxy-acetylene flame which is used for preheating. An oxygen and iron reaction takes place within the flame, liberating a large quantity of heat. (See Fig. 85.)

The iron powder is introduced to the reaction area which is just above the surface of the oxidation resistant material to be cut. A

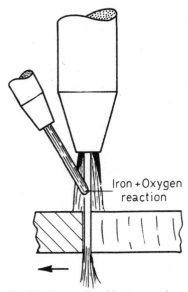

Iron + Oxygen
reaction

Fig. 85. Heat produced by iron powder and oxygen.

suitable pressure of nitrogen or compressed air is used to convey the iron powder to the reaction area. Oxygen should never be used because it may result in a flash-back within the cutting blowpipe with the possibility of serious consequences.

The granular powder usually contains a certain amount of an abrasive substance to erode or wear away edges of the material to be cut. Also, some of the powdered mixture is capable of combining with the material at high temperatures, which has the effect of dissolving or fluxing the oxides away. The high velocity oxygen removes the molten base material as well as the oxidized iron.

Chemistry of the process

The chemical reaction involved, when powder injection cutting, is quite similar to the reaction which occurs when oxy-acetylene cutting steel containing less than 0·3% carbon. The difference, when powder cutting, is that the iron–oxygen reaction takes place above the surface of the material to be cut. The exothermic reaction that is obtained from the combination of iron and oxygen liberates a far greater amount of heat than would be obtained without the addition of the iron powder reaction. The resulting oxidation is on a larger scale than when cutting mild steel with the oxy-acetylene process alone. The desirable result of this is to provide a larger quantity of heat at higher temperatures, thus allowing a larger area of material to be heated.

When non-ferrous materials are cut by this process, most of the material removed is done by a melting and eroding action taking place throughout the material. The high velocity oxygen cutting stream completes the operation by removal of the molten products in its path.

Fuel gas used
The preferred preheating fuel gas, when powder injection cutting, is acetylene, because of its high intensity of heat. The purpose of the preheat flames is as follows:
  (1) To raise the temperature of the iron powder in the reaction area to its ignition temperature so that the iron–oxygen reaction may take place.
  (2) To ensure, when cutting is taking place, that a correct preheat level is maintained at the edges to be cut. It is also necessary to overcome any heat losses by conduction.
  (3) The combustion of acetylene provides reducing gases which protect the cut edges from oxidation during cooling.

Materials which may be cut by this process

The process lends itself to the severing of a variety of materials. The quality of the cut varies from excellent, with certain materials, to only fair, with a few materials.

Stainless steel
This is readily cut by this process. The nozzle is held about 1 in to 2 in above the surface to be cut so as to allow the iron powder to enter the reaction area and cause oxidation to occur above the surface to be cut. The heat liberated from the reaction allows 'flying starts' to take place, thus eliminating the need for preheating. Cuts on stainless steels made by this process are of a high quality.

Nickel: nickel alloys
These may be regarded as difficult to cut materials. It appears that pure nickel is more difficult to cut than its alloys. The chemical combination of nickel and oxygen does not readily take place. A limited amount of success has been attained but only on thinner sections.

Cast-iron
Cast-iron may be cut by the conventional oxy-acetylene process, but the quality of the cut is regarded as poor. The powder injection process is capable of producing high quality cuts on cast-iron. Because cast-iron is inherently brittle, preheating is generally carried out to prevent cracking.

Alloy steels (containing alloying elements less than 20%)
Many of the steels in this group have quench hardening tendencies. To prevent the formation of martensite, which is hard and brittle and may lead to cracking, a suitable preheat is required. The resulting cut is of an extremely high quality.

Copper
This metal is an excellent conductor of heat. To offset this, a high preheat is necessary to enable an even balance of heat to be maintained. The quality of the cut on thinner sections can be regarded as average.

Aluminium
The physical properties of aluminium are mainly responsible for the poor quality of the cuts obtained. Aluminium melts at 658°C whilst its oxide melts at over 2 000°C. This means that before the oxide can be successfully removed, the metal will be at a temperature above its own melting point, causing a cut which is ragged.

Oxygen-arc cutting process

Principle
The oxygen-arc cutting process offers the ability to severe a wide range of materials. Examples of these are quite similar to the materials which may be cut by the powder injection process; mild steel, alloy steel up to 20% alloying elements and stainless steels are cut quite readily. Cuts on other materials can only be described as fair. The cut is obtained by oxidation and the melting action of the electrode. (See Fig. 86.)

When cutting is carried out on mild steels, the material is heated to the ignition temperature by means of an electric arc struck between a hollow-coated electrode and the material to be cut; thus the preheat is provided. The preheat to ignition temperature is instantaneous and a jet of pure oxygen is directed through the tubular electrode on to the heated area. Oxidation immediately takes place, liberating a large amount of heat energy. The coated electrode is consumed during cutting, this combines, forming an iron–oxygen reaction, therefore increasing the amount and intensity of heat available. The coating around the electrode assists in maintaining the arc stability and also provides a small crucible at the tip of the electrode. This serves the purpose of directing and confining the heat of the arc and oxidation reaction to one localized spot.

159

When considering the cutting of other materials, the higher intensity and quantity of heat liberated from the arc, added to the iron electrode core metal reaction with oxygen, supplies the preheat source. This source is intense and localized and offers the following advantages:

(*a*) Cutting may start immediately the arc is struck.

(*b*) Some materials, which are oxidation resisting, are readily cut by this process.

(*c*) When compared to oxy-acetylene cutting, the speed is considerably faster.

(*d*) The cutting of heavily rusted and corroded materials may be carried out.

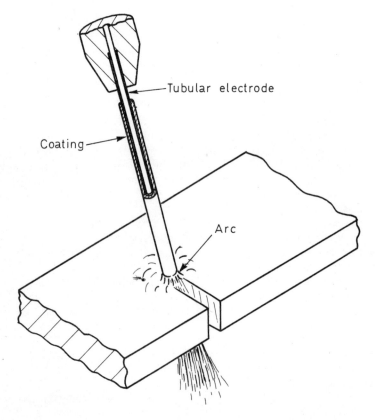

Fig. 86. Oxy-arc cutting.

Effect of cutting on low alloy steels

Because of the high cutting speeds, there is a pronounced quench effect exerted by the conduction of heat into the material being cut. There may be a slight migration of carbon to the heated cut edge, which, in turn, increases the hardness of the cut edge.

The surface of the cut is not quite of the standard of an oxy-acetylene cut made on the same thickness of material. But it should not be inferred that the cut is of a low standard, indeed welding may often be carried out on edges which have been prepared by the oxy-arc cutting process.

Plasma arc cutting

It is possible to cut any metal with this process. The reason for this is that the process does not rely upon a chemical reaction, but instead a thermal condition is responsible for its success.

Principle

The initial heat is created by an electric arc which is struck between a tungsten electrode and the material to be cut. The electrode is connected to a negative terminal of a d.c. supply, the material to be cut being connected to the positive terminal. Therefore two-thirds of the heat generated by the arc is where it is required, i.e., on the material to be cut. The tip of the tungsten electrode is recessed inside a nozzle which has a small diameter orifice. The effect of this is to direct the arc on to a small area on the surface of the material to be cut, providing intense local heat. A gas, or a mixture of gases, is passed through the nozzle and, when in the vicinity of the arc, is heated and therefore rapidly expands. The expanded gas is then forced through the small orifice of the nozzle and so accelerated in transit. It is now possible to state that the process employs a high temperature, high velocity constricted arc. (See Fig. 87.)

When cutting is being carried out, the surface of the material being cut is rapidly preheated by the arc. This preheating, along with the pressure of the high temperature and high speed cutting gases, results in cutting taking place immediately. The cut is carried through the thickness of the material by the high speed, heated gases continually melting and then removing the molten metal; hence it is a thermal action.

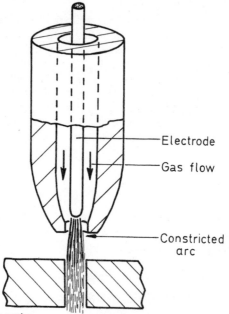

Fig. 87. Plasma arc cutting.

Gases used

The arc is struck in an argon atmosphere, because argon offers less resistance to the formation of an arc than would be the case if hydrogen was used. When the arc has been formed, hydrogen is blended into the gas stream, forming a mixture of argon and hydrogen. Argon is an inert gas, whilst hydrogen is a reducing gas, therefore the cut edges are protected from the atmospheric gases whilst at high temperatures.

A mixture of nitrogen and hydrogen may be used, reducing the cost of the process. But the objection to these gases is the fact that the quality of the cut edge is inferior to the cut edge when argon and hydrogen are used.

For the purpose of cutting mild steel, oxygen can be used in addition to the mixtures already mentioned. The advantage of oxygen is that the iron–oxygen reaction will take place releasing additional heat, also thicker sections may be cut and the rate of cutting is quicker. But there is an objection to the use of oxygen; this is that through oxidation the life of the tungsten electrode is reduced.

Effects of cutting

Generally speaking, the cuts on all materials made by this process are exceptionally good. When compared to other cutting processes there is less alteration to the metallurgical and physical properties of the material which has been cut. This is because the speed of cutting is exceptionally fast. With the exception of carbon steels and cast-iron, there is little alteration in the hardness of the cut edge. The reason for this is that there is no loss of any alloy from the cut edge. The width of the heat-affected zone varies, depending on the thickness and conduction properties of the material being cut.

Laser cutting

The use of lasers in relation to welding and cutting of various materials is becoming important. If a metal or alloy is capable of being melted without immediate decomposition, then it may be severed by the laser beam. The laser directs and amplifies the intensity of a light wave. When this is directed on to a material, rapid melting and evaporation result. Because of evaporation of the material, it is only possible, when welding with this process, to make the welds in the form of overlapping spots. However, when cutting, the rapid evaporation of the material is no problem. This is because the material being cut is removed partially by a melting action and partially by evaporation taking place.

Heat energy

The heat energy required for cutting is obtained by the use of a highly-focused radiation beam. On striking the surface of the material to be cut, the energy of the laser beam is liberated in the form of very intense heat. Most of this heat is used in melting the material, whilst some is consumed during the evaporation of the liquid metal. Very small losses of heat may be accounted to conduction through the material and by reflection from the surface of the material being cut.

Advantages

The laser beam offers several advantages when used as a cutting tool. These include:

(1) Because the beam does not diverge, the distance between the laser head and the material being cut does not matter, therefore awkward position cuts present no problem.

(2) There is no contact between the work and the cutting head.
(3) The piercing of holes in virtually any metal or alloy is a practical proposition.

Unfamiliar terms used in this chapter

Ignition temperature: The temperature at which iron will burn with oxygen.
Velocity: Speed.
Hydrocarbon: A compound composed of hydrogen and carbon.
Migration: In this case, carbon moving towards the heated cut edge.
Erode: To wear away, corrode.
Exothermic: A reaction in which heat is liberated.
Flying starts: Starting to cut without a time period allowed for preheating.
Constricted arc: An arc which is between the work and the tip of a tungsten electrode which is recessed within a nozzle orifice.

Chapter reviewing questions

What is combustion?
How does combustion affect cutting?
Compare the merits of the respective fuel gases.
What are the impurities that may be present in oxygen?
Describe the effect of cutting on the edge of a piece of mild steel.
Discuss the effect of alloying elements on cutting with the oxy-acetylene process.
Why is stainless steel difficult to cut by the oxy-acetylene process?
When oxy-acetylene cutting, what would be the effect of introducing iron powder into the reaction zone?
Can non-ferrous materials be cut with the powder injection process?
What gas is preferred as a fuel gas when powder cutting?
Name the advantages of oxy-arc cutting.
What is the principle of tungsten inert gas cutting?
Why, when plasma arc cutting, is the arc struck in an atmosphere of argon?
How is heat transferred in laser cutting?
When laser cutting, how is the metal removed from the cut area?

Index

Age hardening, 127
Alloying elements, 1–2
Alpha iron, 9
Aluminium: properties, 1–2, 120–2, 131; welding of pure aluminium and non heat-treatable alloys, 122–4, 125; effects of welding upon cold-rolled non heat-treatable alloys, 125–8; welding heat-treatable alloys, 128–30
Aluminium bronze: two types of, 145–7; welding difficulties and processes, 147
Aluminium oxide (alumina), 120, 121
Annealing: steel, 24–5, 43; aluminium, 120; copper, 137
Argon: used in plasma arc cutting process, 162; backing, 77; gas shield, 125, 147
Atmospheric gases, effect on welds, 59–60
Austenite, 10–11, 16–20, 22, 25, 28, 37–42, 46, 58, 62, 63, 71, 72
Austenitic manganese steel, 6, 115–16; hardsurfacing with, 116–18
Austenitic stainless steel, 63; properties, 73–6; welding of, 76–8; weld processes, 78–9; joined to other materials, 99–101; for corrosion-resisting conditions, 101; deposited on base material when hardsurfacing with austenitic manganese steel, 117–18; as hardsurfacing material, 118

Blackheart malleable iron castings, 90, 91
'Body centred cubic' atomic arrangement, 3, 4
Brasses: properties, 141–3; welding

difficulties and processes, 143–4
Brazing: spheroidal graphite iron, 93–4; for joining copper to nickel, 100
Brittle and ductile fractures, 50–51
Bronzewelding: cast-iron, 91–2; for joining dissimilar materials, 100, 101; for copper, 135; for joining tough pitch copper, 139
Buttering layer, 87–8, 97–101

Carbides: formation of, 86; in bronze-weld deposit, 98; result in creep resistance of the alloy, 79; precipitated in hardsurfacing with austenitic manganese steel, 116, 117
Carbon, as alloying element: more influence upon hardening of base material than other elements, 64; in stainless steel, 70–3; in heat resisting steels, 79; in cast-irons, 82–4; in low alloy steels, 110; in cobalt-based alloys, 111, 112; in austenitic manganese steel, 115
Carbon-arc welding process, for hardsurfacing, 108, 110
Carbon dioxide shielded arc welding process, for low alloy steels, 65
Carbon pick-up, in welding austenitic stainless steel, 77
Cartridge brass, 141–2
Cast-iron: 82, 83; malleable, 89–94, 98
Cathodic cleaning action, 123, 125, 132
Cementite (iron carbide or carbide of iron), 11, 17–19, 23, 27, 39; formation of, 12; in cast-irons, 82–4, 90
Chromium, an alloying element: in low alloy steels, 63, 80, 110, 111; in

Index © Cassell & Co. Ltd. 1968